T0219938

Universe Without Things

Jan-Markus Schwindt

Universe Without Things

Physics in an Intangible Reality

 Springer

Jan-Markus Schwindt
Dossenheim, Germany

ISBN 978-3-662-65425-5   ISBN 978-3-662-65426-2   (eBook)
https://doi.org/10.1007/978-3-662-65426-2

Contents

List of Figures

1

Introduction

I think there are basically two different motivations to deal more intensively with physics. One is the **engineering** approach. You want to understand how the technology around you works, the car, the refrigerator, the lamp, the mobile phone, the energy supply; what natural laws underlie it and how you can use these laws to design something new and thus solve practical problems.

The other is the **philosophical** approach. Here one is driven by questions like: What is the world? How did it come into being? Will it end one day? What is the "reality" behind things? Is everything ultimately mathematically comprehensible? And what am I? A lump of atoms? Some kind of computer program running on a brain? Or something else entirely? How are space and time related? And so on.

For me, the motive was clearly philosophical. When I was little, my father used to take me to shows at the Mannheim Planetarium. The stars and planets themselves weren't that important to me; I wanted to know how big the universe was, whether it had a boundary, how it started. I wanted mysterious words like the "big bang", "expansion" and "space-time" explained to me. And because they were mentioned mostly only in passing, I often became impatient.

At 13, I took matters into my own hands, learned special relativity, began devising my own thought experiments and drawing conclusions, most of which were complete nonsense. Feverishly driven by my questions, over the next 2 years I devoured every popular science book on fundamental physics I could get my hands on.

At 16, I lost sight of physics for a while in the turmoil of a typical teenager's life. But at least I remained faithful to mathematics, which had always

© Springer-Verlag GmbH Germany, part of Springer Nature 2022
J.-M. Schwindt, *Universe Without Things*,
https://doi.org/10.1007/978-3-662-65426-2_1

fascinated me in parallel with physics, and successfully participated in national math competitions, through which I also met many like-minded people.

This seething time ended with the Abitur and the question of whether I should study mathematics or physics. Mathematics is poetry, beauty in its purest form. But it is also infinite. Any set of assumptions (axioms) can be chosen as a starting point, and for all eternity one can unroll the infinitely branching chain of conclusions that bubble out of it through logic. *"So if that keyboard is infinite, then there is no music you could play on it,"* says the main character in the story *Novecento* by Baricco (2003, p. 75). This was the feeling that came over me at the time when faced with mathematics. In physics, the number of problems is large but limited. It describes the world in which I live, and therefore concerns the fibers of my existence. So I chose physics because all my questions led me to it.

Because so much of physics is highly relevant to these questions. Here is a first overview of the topics:

- **Reductionism with surprises:** Indeed, the principle of reduction pervades all of natural science. Biology is reduced to chemistry: An organism is understood from the interaction of its organs, the organs from the interaction of the cells, the cells from the molecules and the chemical processes that take place on them. All chemistry in turn follows from the rules of quantum mechanics and statistical physics. There are such hierarchies within physics as well. One phenomenon is reduced to another, equations to other equations, theories to other, more "fundamental" theories. But while at the beginning this chain of reductions still amounts to a continued decomposition and zooming in, suddenly stages occur in which, instead of a decomposition, a complete transformation takes place. Newtonian gravitation suddenly transforms into pure geometry when reduced to general relativity. And the elementary particles seem to live in a completely different world than the objects that are composed of them.

- **Determinism and chance:** Determinism means that the world runs like clockwork. The present state completely determines the state at all later times and in turn follows from the earlier states, without any leeway. Thus, all history is already determined in the initial conditions at the beginning of the universe. If determinism is true. The description that Classical Mechanics gives of the world is indeed deterministic. In Quantum Mechanics, on the other hand, everything seems to be driven by chance. How the two theories relate is extremely complicated. However, quantum mechanics is the more fundamental of the two in terms of reductionism. So chance does rule the world after all? On the other hand, there is also *seeming* randomness, such as the throwing of a die, the outcome of which we cannot predict only because we are unable to analyze the problem quickly and accurately enough (throwing trajectory, rotation, elasticity on impact, etc.). Some say that even the randomness of quantum mechanics is only seeming; there are different interpretations here. The situation is tricky and hard to decide. To make things even more complicated, there is also the subjective feeling of "free will", which allows us to make decisions at least sometimes unimpressed by physical facts. And that doesn't fit very well with either determinism or pure chance.

While physics may not be able to decide this question unequivocally, it is exciting to see what it has to contribute to the topic, because there are some amazing facets to it. After all, physics is the greatest of philosophers; it comes up with ideas that none of its brooding human colleagues would ever have thought of.

- **Relationship of observer, theory and world:** An observercan observe another observer observing (voyeurism has no limits). The scientist therefore takes on a strange dual role. On the one hand, it is he who makes measurements and theories, and in doing so he has the feeling that he is acting as a free being, bound only to his reason and curiosity, perhaps also to his ambition – quasi outside the strictly lawful circumstances he is investigating. He is master of his measuring apparatus, with which he forces nature to answer his question. At the same time, however, he is a component of the world he investigates; he can himself be analyzed and measured, and thus becomes the object of theories. His body consists of the very strange particles with which he conducts experiments. He is himself part of the picture he draws with his theories. This gives rise to all kinds of entanglements, and in no other theory are these expressed so blatantly and clearly as in quantum mechanics, with which we will deal in detail.

  How are observer, world and theory connected? Physical theories represent mathematical structures. So does the world consist of things that are fully described by mathematical structures? Or are things even mathematical structures themselves? If things are "objective," what could possibly be "objective" about them beyond pure mathematics? Are we ourselves, in the end, pure mathematics? Or are theories rather an incomplete representation of the world, consisting of the very aspects that the "scientific method" can deal with? Or is what we call "the world" itself only a theory, a projection of our mind?

Quantum mechanics is the greatest stroke of genius physics has devised. It pokes us right in the head with these questions and forces us not to dismiss them with a banal form of materialism.

- **Cosmology, the evolution of** the **universe:** Cosmology is the branch of physics that deals with the universe as a whole. As such, it tells a story of the universe. It begins about 14 billion years ago with the **Big Bang,** about whose exact course of events we presumably cannot know anything. Our knowledge begins a few fractions of a second later, with a dense hot plasma, an elementary particle soup, from which all further structures emerge step by step, while the universe cools down and expands. After 400,000 years, it becomes transparent and thus bright as the first atoms form from electrons and protons. And so the story continues with great detail and precision, right up to the present day, when our blue planet chugs comfortably around its sun.

- **Relativity of facts:** What actually is a fact? We describe something (an event, a process, properties of an object, etc.) with the help of sentences of our language. In doing so, we insert the event or process into the grammatical structure and vocabulary of our language. How "objective" are such descriptions? Do we, for example, impose a structure on reality with our grammar that it does not actually have? Even in the same language, descriptions of the same process sound very different when they come from different people. But if the description comes from a different language with a completely different grammar, then even its structure is completely different. But often there is an "objective core", something that is expressed in all descriptions, no matter in which language and by which person, and that is just the fact.

  It is similar in physics. One can describe processes in different coordinate systems, in different units of measurement, with different formalisms. But the goal is always to keep the "objective core" in mind, because that is what is actually "physical", what remains after the relative, i.e. context-dependent details have been peeled out. The great merit of Einstein's two theories of relativity (the special and the general one) is the demonstration that we need to peel much further than originally thought. Many of the things we had thought were objective and unambiguous are not at all. Crucially, however, there continues to be *some* objective core, it just lies a bit "deeper" than suspected. Another great merit of these theories is that they provide, on the basis of the remaining core, "interpreter procedures" for translating the contextual details of one perspective into those of another. It is in these unambiguous translations (in physics they are indeed unambiguous!) that the objective core is expressed.

Such "peeling out the core" and translating contexts back and forth can be found in all areas of physics. In most cases, it takes place within *one* theory. However, it becomes even more exciting when a process can be described with different theories and one then has to translate back and forth between the theories, e.g. between Newtonian gravity and general relativity.

- **Time:** What actually is time? Already Augustine answered the question thus: *"If no one asks me about it, I know it, but if I should explain it to one who asks me, I do not know it"* (*Confessiones* XI, 14). Time is as ubiquitous in physics as it is in life. However, there it seems to have such a wholly different character from what we know. The theories of relativity show how space and time are interwoven into a common four-dimensional geometric entity, **spacetime**. They further demonstrate to us how the amount of time that lies between two events depends on the perspective of the observer, namely where he is and at what speed he is moving. The term "simultaneous" loses its absolute meaning.

  Moreover, at the microscopic level, physical theories are **time-reversal invariant**, i.e., any process described by them can proceed both forward and backward; one can simply reverse time. So there is no clear causality at this level: before and after, cause and effect are interchangeable. Now this completely contradicts our experience. How can this contradiction be resolved? It turns out that this is possible, but quite complicated, and has to do with the concept of **entropy**.

  In this topic, too, physics thus brings to light unexpected facets and illuminates the subject from many different sides. In the end, however, we will be left with the impression that, despite all this, physics cannot fully get to the bottom of time. That certain aspects of how we experience time simply have no counterpart in physics.

This overview gives a first hint of how much physics has to say about fundamental philosophical problems. Whoever is driven by such questions cannot avoid physics. Many of these questions cannot be answered unambiguously by physics. (Research optimists would like to add a "yet" here.) It does, however, elevate our not knowing to a much higher level. It shows us facets of the problem in question that we would never have expected. This is perhaps already the highest that philosophy can achieve.

It was only late that I understood how important it is to deal with experiments. In physics studies there are lectures on theoretical and experimental physics, and a few experimental practical courses. Tending to come from mathematics and philosophy, I wasn't interested in all the experimental techniques, I wanted to understand the theories that ultimately resulted from the countless experiments that I simply took for granted. Experimental physics does not deserve this disregard. Experiments are the foundation and whole substance of physics. A theory without experiment is pure mathematics. It is only through experiment that one understands how the knowledge that feeds a theory is arrived at. And only through experiment is it defined what the mathematical symbols in the theories actually mean, what they refer to.

Theory without experiment is a major problem in fundamental physics today, and at the same time a consequence of its success. For so well do the known theories work in the areas of the universe accessible to us that most of the great open questions take place in a territory inaccessible to our experimentation.

I was also affected by this in my research. In my diploma thesis I still dealt with "tangible things": with the "clumping" of matter in the early universe, which led to the formation of galaxies and galaxy clusters. There were detailed observational data on this, and with these the possible values of parameters in our theories could be narrowed down. After that, however, I was drawn further and further into more fundamental questions, into speculative areas such as "extra dimensions" and "quantum gravity". I did my PhD and research for another 4 years as a postdoc, but more and more I had doubts whether I was on the right track. I loved the teaching tasks, working with students, and with pleasure I gave a lecture on general relativity. But the research became more and more alien to me. That was the paradox: the more I approached thematically the areas that seemed fundamental for understanding reality, the more this very work seemed to lose touch with reality. Something had gone wrong.

In the tightly timed academic life with its temporary positions, there is no time for temporary retreat. As soon as you have taken up a new job, you already have to apply for the next one, you have to distinguish yourself through research work; there's nothing to be gained from teaching alone. Having become disoriented as a researcher, I decided to go into exile in the IT industry. But the big questions continued to burn.

I also knew that I was not alone in my frustration. The view that research in fundamental physics is in crisis is shared by many. From the nineteenth century until the 1970s, there was a non-stop fireworks of discoveries; theory and experiment went hand in hand, one spurring the other. For 40 years now, however, progress has slowed considerably. The 1973 **Standard Model of particle physics** is just too good; together with the **Standard Model of**

cosmology, it allows the classification and description of almost all phenomena accessible to us, from the smallest to the largest scales.

This does not mean that the big questions have all been solved. On the contrary, they have only wandered further and further into areas to which we have no access. Thus began an era of speculative theoretical physics not backed up by experiments. The experimental data increased in the same time by numerous technical improvements in gigantic scale, but they contributed quasi nothing to the speculative new theories, but only confirmed over and over again the two standard models.

How can this situation be overcome? Various approaches have been discussed. The influential physicist Lee Smolin of the Perimeter Institute in Canada, for example, argues that[1] the scientific establishment is too attached to the mainstream; the pursuit of "revolutionary ideas", as Einstein had them, is not promoted enough, or even virtually prevented. In Smolin's view, great insights lurk around the next corner, answers to the great remaining questions of physics. All that is needed is to keep researchers with the right lateral thinking traits in line, and then the next scientific revolution will take place.

I think this thesis is implausible. It is not for lack of "revolutionary ideas". There have been a few of those in recent decades, but unfortunately they have not led to success. I think the slowdown in progress is not so much caused by a misdirection of the research enterprise, but rather a consequence of the natural limitations we face in doing physics as a science.

These limits can be divided into practical and fundamental ones. The practical limits are that we cannot make all the observations we would like to make. Since neither the speed of light nor the age of the universe is infinite, we cannot look as far as we like. There is a horizon in the universe beyond which we cannot see. Our energy supplies are limited, so we can't shoot particles at each other with arbitrarily high energy to see what happens.

The fundamental limits are that, in general, not all fundamental questions can be answered with physics. One of these questions is: Why does something exist at all, and not nothing? Physics can describe a universe with its laws, but why this universe should exist at all, it cannot tell us (unless the universe is "born" out of another universe; such theories exist, but this only shifts the problem). On other topics, the fundamental limits are much more controversial, such as our consciousness and the connection between "mind and matter." But before we are ready for that topic, we need to spend a few more chapters on what physics actually is and does.

---

[1] E.g., in Smolin (2006).

So if these limits do exist, and if there is some evidence that we have already come very close to them, then perhaps this is not a crisis at all, but a natural phasing out? This is the thesis put forward by John Horgan (1996) in his book *The End of Science*, and it has drawn a lot of ire. No wonder, who likes to be told that his discipline is coming to an end. But what does "end" mean? Physics is a magnificent castle of great beauty, filled with treasures that can be explored in any of its countless rooms, by any visitor who cares to look. And by no means has every gem in all the chests been duly appreciated. The only ones who have reached their limits are the builders, who are always trying to add another tower on top.

Academic research tries to push the boundaries of **global knowledge**, that is, to find something that no one else has found before you. Meticulous records are kept of who was the first to say what, when, by whom an idea was first expressed, who was the first to carry out an experiment, and it is to that person that all subsequent work on the same subject must point. To survive in academia, to build a career, you have to try, again, to find something new, building on all the previous news, and publish it quickly and first. These are - speaking in the image from above - the builders who want or need to push the boundaries of the castle further and further out, into the sky, by building more and more new towers, one on top of the other. This went very well until about 40 years ago, and served the splendour of the castle. But now thousands of theoretical physicists stand on the roofs at lofty heights, wanting or needing to push on and on, even though the new towers are getting shakier and shakier.

On the other hand, there is also private research. Driven by certain questions, one wants to understand something. All the better if others have already understood it before you, because it's not about pushing the boundaries of global, but of **personal knowledge.** So one delves into a subject, building on what others have written, and grapples with it. To understand and wonder, not to invent a "revolutionary" new idea. In the picture from above, this is the visitor of the castle, who wanders through the rooms, looks at them closely; an object holds him, he looks at it closely, from all sides, he delves into it in wonder, and the impressions gained settle in his mind.

This path is still open even if the thesis of the "phasing out" should be confirmed. This book is therefore intended, among other things, to be a kind of treasure map, giving a description (albeit very incomplete) of which jewels can be found in which rooms of the castle.

However, a popular science book like this can only be a treasure map. You have to dig up the treasures yourself. Physics is written in the language of mathematics, and its findings are based on thousands of experiments. All of this cannot be expressed in a few words in everyday language. If you want to know what it's really all about, you have to dig through the mathematics and the descriptions of the experiments. You have to perform complicated calculations and thought experiments on your own. Just like you don't learn to play a musical instrument by reading a few books.

2

Philosophy

Amazement is at the beginning of all philosophy, writes Plato. We marvel at the world and at ourselves. We ask ourselves where it all comes from and whether and how it will end one day. We realize that things, and we ourselves, are subject to constant change and wonder if there is a constant basis for it all, a "substance" (or substances) that makes it all up, that takes on new forms all the time but remains the same in itself. We see that we are often deceived by the appearance of things, we doubt and ask what is real.

We understand that we are mortal, that we do not have much time, and we ask ourselves how to live in the face of this. What is the "meaning" of life. And whether there is something beyond life for us, on the other side of death, or outside of time.

This questioning, doubting and wondering is inherent in all of us. As children, at a certain age, we bombard our parents with questions, wanting to know and understand everything. For some, however, the great wonder wears off over time; we become too absorbed by the worries of everyday life and the pursuit of goals that are no longer questioned. For others, however, it continues, accompanying and shaping life and thought to the end. These are the "philosophical natures" among people.

Also the others, the non-philosophers, have a certain picture of the world and of man, they are just not so clearly aware of it, or they take it as something given, do not question it, or only very superficially. Yet this picture is crucial for our lives. In the words of Karl Jaspers:

© Springer-Verlag GmbH Germany, part of Springer Nature 2022
J.-M. Schwindt, *Universe Without Things*,
https://doi.org/10.1007/978-3-662-65426-2_2

For the picture of man which we hold to be true becomes itself a factor in our life. It determines the ways in which we deal with ourselves and with our fellow human beings, our mood in life and our choice of tasks. (Jaspers 1960, p. 151)

The non-philosophers are right and wrong. They are right because philosophizing requires overcoming fears, because it involves effort, and because it sometimes brings before our eyes the abyss (examples follow below) that we have been facing all along, but of which we had no inkling. Philosophizing confronts the certainty of death and one's own transgressions and errors, it tests everything and literally goes to our substance, confronts us with the questionability, the limitedness of all knowledge and cognition. Moreover, there is no end to questioning. At the end of the path there are no clear answers, no "Oh, that's how it is", but only ever new questions and doubts; everything seems to go round in circles.

But they are also wrong, because suppressing and ignoring them does not change the fact that the questions are simply there. Philosophizing does not reward us with clear answers, but it sets a process in motion within us that opens up new horizons, trains our thinking in a different way than everyday problems or mathematics do, and also loosens up something in us – "mental hardenings" is what I will call it – and frees us from prejudices. The process changes our view of the world and ourselves and is capable of changing our lives.

2.1   Philosophy and Language

In philosophizing, the questions that drive us often wash strange thoughts onto the shore of our minds. Thoughts that are not so easily put into words and conveyed, that thus require us to expand our means of linguistic expression, our linguistic "tools of the trade". This is why philosophy is so often a struggle with language. Countless word creations go back to it.

Literature is thus by nature very close to philosophy. It too, in its higher forms, wrestles with language, brings forth new forms, links, metaphors in it, thus opening up new perspectives. It too penetrates into human destinies and feelings, in ordinary and in extreme situations, and tries to draw general conclusions from them ("the moral of the story," in the simplest case) or to clarify, to bring before our eyes peculiarities, things worth thinking about. Every great novel not only tells a story, but deals with the questions of life. Therefore, every great novel is also a piece of philosophy. To take a few examples: in *Anna Karenina*, Tolstoy deals with the nature and purpose of family and marriage;

War and Peace ends with a treatise on the philosophy of history and the entanglement of the will of the individual with world events. Max Frisch is concerned above all with the problem of our individual identity: *"Sooner or later every man invents a story for himself, which he believes to be his life,"* he says in *Mein Name sei Gantenbein*.[1] He describes the roles we must constantly play in order to maintain the image of ourselves (our own and that of others), the freedom we thereby give up.

There is a passage from the novel *A heart so white* by Javier Marias that I would like to quote in full here because it illustrates what good philosophy is in my eyes, and because it still touches and confuses me even after the 100th reading:

> *And I was impatient because I was aware that what I didn't hear now I never would hear; there would be no instant replay, as there can be when you listen to a tape or watch a video and can press the rewind button, rather, any whisper not apprehended or understood there and then would be lost forever. That's the unfortunate thing about what happens to us and remains unrecorded, or worse still, unknow or unseen or unheard, for later, there's no way it can be recovered. The day we didn't spend together we will never have spent together, what someone was going say to us on the phone when they called and we didn't answer will never be said, at least not exactly the same thing said in exactly the same spirit; and everything will be sightly different or even completely different because of that lack of courage which dissuades us from talking to you. But even if we were together that day or at home when that person phoned or we dared to speak to them, overcoming our fear and forgetting the risks involved, even then, none of that will ever be repeated and consequently a time will come when having been together will be the same as not having been together, and having picked up the phone the same as not having done so, and having dared to speak to you the same as if we'd remained silent. Even the most indelible things are of fixed duration, just like the things that leave no trace or never even happen, and if we're far-sighted enough to note down or record or film those things, and accumulate loads of souvenirs and mementos and even try to replace what has happened by a simple note or record or statement, so that, right from the start, what in fact happens are our notes or our recordings or our films and nothing more, even in that infinite perfecting of repetition we will have lost the time in which those events actually took place (even if it were only the time it took to note them down) and while we try to relive it or reproduce it or make it come back and prevent it becoming the past, another different time will be happening, and in that other time we will doubtless not be together, we will pick up no phones, we will not dare to do anything, unable to prevent any crime or death (on the other hand, we won't commit any or cause any) because, in our morbid attempt to prevent time from ending, to cause*

---

[1] Frisch (1964).

what is over to return, we will be letting that other time slip past us as if it were not ours.

Thus what we see and hear comes to be similar and even the same as what we didn't see or hear, it's just a question of time, or of our own disappearance. And, despite that, we cannot stop focusing our lives on hearing and seeing and witnessing and knowing, in the belief that these lives of ours depend on our spending a day together or answering a phone call or daring to do something or committing a crime or causing a death and knowing that that was how it was. Sometimes I have the feeling that nothing that happens happens, because nothing happens without interruption, nothing lasts or endures or is ceaselessly remembered, and even the most monotonous and routine of existences, by its apparent repetitiveness, gradually cancels itself out, negates itself, until nothing is anything and no one is anyone they were before, and the weak wheel of the world is pushed along by forgetful beings who hear and see and know what is not said, never happens, is unknowable and unverifiable. What takes place is identical to what doesn't take place, what we dismiss or allow to slip by us is identical to what we accept and seize, what we experience identical to what we never try, and yet we spend our lives in a process of choosing and rejecting and selecting, in drawing a line to separate these identical things and make of our story a unique story that we can remember and that can be told. We pour all our intelligence and our feelings and our enthusiasm into the task of discriminating between things that will all be made equal, if they haven't already been, and that's why we're so full of regrets and lost opportunities, or confirmations and reaffirmations and opportunities grasped, when the truth is that nothing is affirmed and everything is constantly in the process of being lost. Or perhaps there never was anything. (Marias 1995, p. 38 ff.)

This philosophical "eruption" in the middle of the novel deals with the theme of time and transience. It is not just a theoretical discussion, but literally about our lives. The artful use of language, the metaphors and examples surround the problem with a "soft" logic, a logic that can express much more in this area than the "hard" logic of argumentative technical discussion. This element of expression is, in my view, crucial to good philosophy. You can literally feel the author's struggle with the subject in it. It drives him to the abyss that opens up in the quoted passage: *"Sometimes I have the feeling that nothing that happens happens"* – *"Or perhaps there never was anything"*.

This is what I meant earlier by abyss. The realization of the omnipotence of transience drives the narrator further inward to a point where there is a doubt that goes beyond what seemed possible from a normal everyday thinking, a doubt that suddenly opens up in the black depths before one's thought, a step beyond logic, therefore more feeling than thought. *"Sometimes I have the feeling that nothing that happens happens"* – *"Or perhaps there never was anything"*.

The inexhaustible subject of time will continue to occupy us in this book. We will see that physics sheds new light on some aspects of time and thus has something to contribute to the richness of the subject; but also that there are aspects of time to which physics has nothing to say, puzzles which it is unable to solve because they lie beyond the fundamental limits of physics.

Philosophy often expresses itself in the form of contradictions, paradoxes (*"nothing that happens happens"*). For logic only leads us forward, starting from a chosen set of concepts and assumptions that we form from these concepts, in baby steps, without bringing anything new into play. It shows us what the conceptual system and the assumptions *imply*, illuminates, so to speak, the world drawn by them. But to really move forward, it is often precisely crucial to escape the limits of the conceptual system and assumptions, to get beyond them. By the standards of logic, the sentence *"nothing that happens happens"* is nonsensical, but related to the experience of the narrator, it makes sense, expresses something that logic is not able to express, at least not in the terms of everyday language.

This is true, by the way, not only in philosophy but also in the natural sciences. In both cases, in contrast to mathematics, it is true that concepts are not considered only for themselves, according to their logical connections, through which they are defined, but refer to something that lies outside of themselves, in our experience, even if possibly very indirectly. The sentence "The electron is both a wave and a particle" does not make the slightest sense within the framework of classical physics, expresses a contradiction. Yet there are plenty of experiments, things of our experience that is, whose results are best expressed as "The electron is both a wave and a particle." (Of course, again, it requires some steps and terminology to summarize the direct experience of the experimenter during the experiments in this sentence). The conceptual system of classical physics is thus broken up, must be extended, even radically modified, to quantum mechanics namely, in which the proposition becomes possible and even finds an explanatory framework.

In philosophy, things are less clear than in the natural sciences. This makes it all the more important to be aware of the limits of a chosen conceptual system; to express things that go beyond such a system; to create new concepts that are capable of expressing something new.

2.2   World and Mind: Schrödinger's Dilemma

But even new concepts are often incapable of simply resolving fundamental contradictions that we become aware of in philosophizing, and so it is already an important step to *express* them in the given language. A great piece of philosophy that expresses such a fundamental contradiction comes from the physicist Erwin Schrödinger:

> *The thing that bewilders us is the curious double role that the conscious mind acquires. On the one hand it is the stage, and the only stage on which this whole world-process takes place, or the vessel or container that contains it all and outside which there is nothing. On the other hand we gather the impression, maybe the deceptive impression, that within this world-bustle the conscious mind is tied up with certain very particular organs (brains), which while doubtless the most interesting contraption in animal and plant physiology are yet not unique, not sui generis; for like so many others they serve after all only to maintain the lives of their owners, and it is only to this that they owe their having been elaborated in the process of speciation by natural selection.*
>
> *Sometimes a painter introduces into his large picture, or a poet into his long poem, an unpretending subordinate character who is himself. Thus the poet of the Odyssey has, I suppose, meant himself by the blind bard who in the hall of the Phaeacians sings about the battles of Troy and moves the battered hero to tears. In the same way we meet in the song of the Nibelungs, when they traverse the Austrian lands, with a poet who is suspected to be the author of the whole epic. In Dürer's All-Saints picture two circles of believers are gathered in prayer around the Trinity high up in the skies, a circle of the blessed above, and a circle of humans on the earth. Among the latter are kings and emperors and popes, but also, if I am not mistaken, the portrait of the artist himself, as a humble side-figure that might as well be missing.*
>
> *To me this seems to be the best simile of the bewildering double role of mind. On the one hand mind is the artist who has produced the whole; in the accomplished work, however, it is but an insignificant accessory that might be absent without detracting from the total effect.* (Schrödinger 1958, reprinted in Dürr 1986, p. 167 f.)

The contrast between the world in the mind and the mind in the world, which Schrödinger describes here, is truly fundamental. In my view, it is the core problem of all philosophy. It reminds me of Escher's picture "Drawing Hands" which shows two hands drawing each other. Imagine the hands labeled "mind" and "world" respectively.

What I call the world is indeed first of all a construct of my mind. It is an extrapolation and generalization of my perceptions and memories, trusting much of what I am told or what I happen to hear or read. I've never been to Australia, but I trust my atlas that the continent exists, and my friends that they've actually been there. I even believe in the Higgs particle, although very few people have seen it, and in a very indirect way. I believe that the sun will rise again tomorrow, although no one can prove that, but I trust that the laws of nature of today will still apply tomorrow and that "things" will still be there. I also believe in my memories, that is, that the past they tell me about actually happened. I believe in them even though it is well known that memories can be deceiving. At least that's what I remember someone saying. Santa Claus, on the other hand, I don't believe in, though someone once told me he existed and never explicitly retracted that claim. We seldom realize how great the discrepancy is between what we take to be certain of the "world" and what we are actually given at any given moment. How much faith, trust, inference, generalization, intuition is involved. Yes, the world (or *my* world) is a construct of my mind. The existence of brains is a part of that construct. And now my mind, which is the "constructor" after all, is supposed to depend in turn on this one part of its construct, to be a kind of software running on it? There is indeed something confusing in this double role of the mind.

In the *Critique of Pure Reason*, Kant also never tires of emphasizing that space and matter only *appear* as something outside of us, but are in fact ideas *within* us, in particular, *"that it is not the motion of matter that causes representations in us, but that motion itself (hence also the matter that makes itself knowable through it) is a mere representation".*[2] Further he writes:

> *So little fear remains that if one took matter away then all thinking and even the existence of thinking beings would be abolished, that it rather shows clearly that if I were to take away the thinking subject, the whole corporeal world would have to disappear, as this is nothing but the appearance in the sensibility of our subject and one mode of its representations.* (Kant 1998, p. 433)

Kant and Schrödinger thus give preference to the world in the mind over the mind in the world. In Schrödinger's case, this is also evident from the fact that he describes the impression that consciousness is bound to brains as "perhaps erroneous", whereas for him there is not the slightest doubt that it is the arena in which all world events take place.

Today, this view seems somewhat alien to us. We have become so accustomed to the objective, scientific view of the world that it seems self-evident to most that primarily the mind is in the world (bound to brains) and only

---

[2] Kant (1998), p. 435.

secondarily, as an image, the world in the mind. This image seems to us to correspond reasonably well to the external actual conditions, especially where it has been purified by scientific thinking. Brain research has also made great progress in the analysis of *how* the brain accomplishes certain mental processes.

Also the philosophy of mind, which wants to illuminate the connection between mind and brain from a philosophical point of view, has been arguing for decades almost exclusively from a perspective in which the material world is the primary, presupposed thing and the task is to derive certain aspects of the mental from it, or to admit that this is not possible and that something must therefore be added to the material.

However, when you get to the bottom of matter, and this is what happens in physics, it turns out that things are not so clear and you have to look very carefully at the relationships between the world and the mind before you ascribe so clearly all power and reality to matter. That is also what this book is about.

In any case, Schrödinger's dual role of the mind becomes relevant when we examine the workings of the mind itself, that is, when the mind is both subject (the "artist") and object (the "image") of inquiry. Here we can often witness an argument, or even an entire line of inquiry, "biting its own tail." An example is the often made argument that our conclusions about the world are almost certainly correct in most cases because natural selection has led us to think correctly, otherwise we would not have survived. But this is circular reasoning, and therefore a fallacy: the theory of natural selection is, after all, itself one of the results of our conclusions; so the argument already presupposes what it is trying to prove. Natural selection and our other scientific results may well be correct. But if I cast doubt on one and seek first to prove it, I cannot presuppose the other as certain. I must also doubt natural selection, because that is *part of* the worldview that results from our scientific conclusions about the world. You can't prove by scientific means that the way science is going is the right way.

In the philosophy of language, language is used to investigate how language works. In other fields, one tries to get insight into the nature of insight, or to get to the bottom of thought by thinking, or to prove the omnipotence of logic by logical argument. All these areas are treacherous, thoroughly prone to circular reasoning and a constant biting of one's own tail. However, the self-judging of the mind is an essential part of questioning and wondering in philosophy. One cannot, therefore, simply ignore these areas. One must learn to walk through this minefield and, where necessary, accept the fundamental limits that are found.

2.3   Devaluation of Philosophy

Despite its paramount importance to our lives, philosophy has been devalued over the last 100 or 200 years, especially among natural scientists. People say "that's purely philosophical" when a consideration is completely irrelevant to the solution of a problem. The "unworldly" intellectual contortions of philosophy are contrasted with the hard facts of the natural sciences, according to a common view. A student of philosophy has to hear from many sides why he does not rather study something "meaningful", "useful". "Nothing will come of it" is the harshest prejudice. Gone are the days when philosophers enjoyed the highest prestige, when the great Aristotle was chosen as the teacher of the great Alexander, the philosopher as teacher and educator for kings. How is it that philosophy has undergone such a decline, at least in the perception of many people, especially most natural scientists? There are several reasons:

- More than two millennia of metaphysical speculations have not led to any definite results. All the alleged proofs concerning the existence of God, the immortality of the soul, the finiteness or infinity of the universe etc. could be neutralized by counter-arguments. Where something could be said about the nature of the world, it was the natural sciences and not metaphysics that delivered verifiable and therefore generally accepted results. The errors of metaphysical speculation were impressively dissected by Kant in his *Critique of Pure Reason*. In it he finally scoffs:

  These sophistical assertions thus open up a dialectical battlefield, where each party will keep the upper hand as long as it is allowed to attack, and will certainly defeat that which is compelled to conduct itself merely defensively. Hence hardy knights [...] are certain of carrying away the laurels of victory if only they take care to have the prerogative of making the last attack, and are not bound to resist a new assault from the opponent. [...] Perhaps after they have exhausted rather than injured each other, they will see on their own that their dispute is nugatory, and part as good friends. (Kant 1998, p. 468)

- Natural scientists tend to see philosophers as their precursors, not as constant partners in the discussion of their results and methods. Natural science, according to the prevailing view, had indeed developed out of philosophy (at least philosophy had played a significant part in its emergence), but then emancipated itself from it, developed its methods, which were superior for the acquisition of knowledge, and naturally disintegrated into numerous individual disciplines. Each of these disciplines was itself responsible for the choice of its methods and the evaluation of its results. The results of all disciplines together would result in our scientific view of the world. There is nothing left for philosophy to do.

- The scientific worldview is considered by many to be a complete description of reality. Underlying this idea is itself a philosophical, even metaphysical position, namely **naturalism,** which states that the "world" exists outside and independently of us and is the reality from which all phenomena and also we ourselves can be explained. The successes of natural science (with the exception of quantum physics) seem to confirm this position.

- The results of physics are so complicated and so deeply rooted in advanced areas of mathematics that philosophical analysis seems impossible without in-depth study of mathematics and physics. Therefore, philosophers, who usually do not have such deep knowledge, are not trusted to have anything meaningful to contribute to the subject.

- On the other hand, the results of quantum mechanics and quantum field theory are also very confusing for physicists, as soon as one wants to derive from them not only statistical predictions for the outcome of certain experiments, but a *world view,* a *meaning* that emerges from them. After decades of fruitless exegesis, most have given up and don't bother with the problem any further. "Shut up and calculate" is a common motto, an oft-quoted saying of the physicist David Mermin, which many understand quite positively, in the sense of Wittgenstein's *"What we cannot speak about we must pass over in silence"*.[3] Again, this is ultimately based on a certain philosophical position, **instrumentalism,** which states that physical theories do not make statements about reality, but are merely tools with which predictions can be derived. Instrumentalism, however, is in contradiction to naturalism, which understands scientific theories very well as statements about reality, the "world"; so there is already a split, a schism, going through the community of natural scientists here.

In the first half of the twentieth century, during the great upheaval with the two theories of relativity, quantum mechanics and nuclear physics, almost every great physicist also left philosophical texts: Einstein, Heisenberg, Schrödinger, Bohr, Eddington, Planck, Born, Pauli and many others. For a compilation of such texts, see Dürr (1986). Today, with the prevailing "shut up and calculate" mentality, any physicist who expresses himself philosophically quickly comes under general suspicion of being a *crackpot,* a crank.

---

[3] Wittgenstein (1963), p. 115.

- This age, dominated by technical achievements, is so successfully concerned with practical problems, is so dominated by "doers", that not much importance is given to the more intellectual mode of philosophy. The science faculties of the universities, too, are now to a great extent allied with industry, are dependent on its funds. As a result, purely academic and thus philosophical interests (at least it seems to most that philosophical interests are purely academic) have been pushed further into the background.

- In addition, philosophers themselves have been calling for the end of philosophy for quite some time. There are several points of view to consider here. On the one hand, many a philosopher believed that he had finally solved all problems and thus brought philosophy to a close. Many philosophers tend to condense their thoughts and views into a "system" which they then believe to be the whole truth. In *Der Mann ohne Eigenschaften*, Robert Musil ironically writes, *"Philosophers are violent criminals who have no army at their disposal and therefore subdue the world by locking it into a system."*[4]

A particularly interesting case is Wittgenstein. In the preface to his famous *Tractatus logico-philosophicus* he writes in 1918:

I therefore believe myself to have found, on all essential points, the final solution of the problems. And if I am not mistaken in this belief, then the second thing in which the value of this work consists is that it shows how little is achieved when these problems are solved. (Wittgenstein 1963, p. 8)[5]

This leads us to the second point of view. Disenchanted by metaphysics and superseded by physics, philosophy found itself severely limited in its sphere of competence, and the problems it had to solve seemed to be of limited scope. This is also how Hawking comments on Wittgenstein in the *Brief History of Time*:

Philosophers reduced the scope of their inquiries so much that Wittgenstein, the most famous philosopher of this century, said, "The sole remaining task for philosophy is the analysis of language." What a comedown from the great tradition of philosophy from Aristotle to Kant! (Hawking 1988)

---

[4] Musil (1952), p. 253.
[5] The year here refers to the German edition.

2.4   Wittgenstein vs. Jaspers

We must indeed concern ourselves a little more with Wittgenstein and his *Tractatus*. Strongly influenced by the logical-mathematical approach to philosophy as exemplified to him by his colleagues Frege and Russell, he wanted the world to be understood in terms of unambiguous, true statements within the framework of a strictly defined conceptual system. *"The world is all that is the case,"* the *Tractatus* begins, *"The world is the totality of facts, not of things."* However, in the course of this book, we will see that facts in physics are a matter just as complicated as things. Crucial to the *Tractatus'* understanding of philosophy is that everything must be clear, logical, and unambiguous, as in mathematics. *"Logic is not a body of doctrine, but a mirror-image of the world".*[6] In language as we use it, however, everything is often not clear, logical, and unambiguous. But this, according to the logic of the *Tractatus*, is then a problem of language, or of its wrong or bad use. For *"what can be said at all can be said clearly".*[7] Wittgenstein concludes that the only thing that then remains are the propositions of the natural sciences. Thus he obtains an arguably truncated philosophy:

The correct method in philosophy would really be the following: to say nothing except what can be said, i.e. propositions of natural science, i.e. something that has nothing to do with philosophy, and then, whenever someone else wanted to say something metaphysical, to demonstrate to him that he had failed to give a meaning to certain signs in his propositions. Although it would not be satisfying to the other person – he would not have the feeling that we were teaching him philosophy – this method would be the only strictly correct one. (Wittgenstein 1963, p. 115)

Wittgenstein does not at all deny that something exists beyond the purview of the natural sciences:

The whole modern conception of the world is founded on the illusion that the so-called laws of nature are the explanations of natural phenomena [...] We feel that even when all possible scientific questions have been answered, the problems of life remain completely untouched. (Wittgenstein 1963, pp. 110, 114)

The laws of nature only describe *how* the world is, but they do not explain anything (we will discuss this later in the book). Within the world described by these laws there is no meaning and no value. Meaning and value are beyond

---

[6] Wittgenstein (1963), p. 101.

[7] Wittgenstein (1963), p. 7.

it. But this "beyond it" belongs, according to Wittgenstein, to the ineffable, the mystical, as he calls it. One simply cannot speak about it, nor even formulate meaningful questions about it. This renders all further philosophy meaningless. *"What we cannot speak about, we must pass over in silence."*

Wittgenstein's criticism does not come from nowhere, in view of the already mentioned fruitlessness of metaphysical speculations. In fact, as discussed above, Kant has already dispelled the errors of metaphysical disputes. The crucial question is whether a compromise, an intermediate solution can be found between the untenable argumentation of metaphysics and the total speechlessness imposed by Wittgenstein. For if Wittgenstein is right, then one can indeed only advise every person tormented by philosophical questions to realize the nonsensical nature of his investigations, to give up, and to refrain from philosophizing afterwards.

The problem with the *Tractatus view*, as I see it, is the strictly logical view that wants to break everything down into true and false propositions. As already mentioned, logic, starting from *given assumptions* and based on a *given conceptual system,* leads us on to the conclusions that can be derived from them. But it says nothing about how to choose the assumptions and the conceptual system. In mathematics, assumptions and concepts can be freely chosen (we will see this in detail in the next chapter); the mathematician can be guided in his choice by aesthetic or practical considerations. But in philosophy we are concerned with our conception of the world and of ourselves; the terms and assumptions relate to something and must be chosen accurately and appropriately. In the natural sciences there are certain procedures, methods, which serve as guidelines for the formation of concepts and assumptions. These then themselves result in something mathematical, a theory, which may satisfy Wittgenstein's strict requirements. But in philosophy, things are less clear. The concepts and assumptions must somehow come "from ourselves." Logic can help us to discard nonsensical conceptual systems and assumptions, but the meaning we give to concepts (unless they are purely mathematical or scientific) is prior to any logic, and cannot be expressed even with unambiguous definitions. There seems to me to be a "felt," "experienced" element here that determines our relationship to "truth," to "existence." If we try to express this felt, experienced element in language, logic can serve us in tidying up, but it does not play the comprehensive role that Russell, Wittgenstein and others give it.

From the beginning, however, philosophy has defined itself in terms of *positions*, statements that were considered true or false, in always somewhat ambiguous metaphysical conceptual systems within which one somehow took a position. Each position was marked with an *ism.* Thus there is monism, dualism, materialism, physicalism, functionalism, epiphenomenalism, naturalism, instrumentalism, theism, deism, atheism, positivism, idealism, solipsism, realism, and so on, there are hundreds of them. Since the positions were mostly understood as *logical* positions, it was a matter of who was right, and so true battles unfolded, fought with logical arguments, just as Kant described the "dialectical battlefield" above; and so it still happens today.

I have the strong impression that the analytical mind, pure reasoning, is not sufficient to take or even define a philosophical position in a meaningful way. I think there are aspects of reality that cannot be fully conceptualized and cast into unambiguous propositions, but can still be thought about and also communicated with a somewhat less rigorous, "softer" logic, such as in Marias' passage. Without considering these aspects, the position lacks something that anchors it in reality, and not just in logic. (Of course, one could say that this view is itself a philosophical position). On the other hand, anyone who wants to do philosophy like mathematics may indeed end up not getting beyond "language criticism" without getting caught up in metaphysical nonsense.

A counter view of what philosophy is (or should be, or can be) has been given by Karl Jaspers. I like his view very much, it represents a compromise between metaphysics and speechlessness, as I wished above. It is also more in line with the understanding of philosophy as I described it at the beginning of the chapter. Let us hear what he has to say:

> *There is a thinking that is not compelling and universal in the sense of science, that therefore has no results that endure as such in forms of knowability. This thinking, which we call philosophical thinking, brings me to myself, has consequences through the inner action carried out with it, awakens the origins in me which also give science its meaning in the first place.* (Jaspers 1960, p. 43)

Jaspers distinguishes between the universally valid truth, whose correctness we can prove, and the truth by which we live. The former is the field of the sciences, the latter the field of philosophy. Truth by which we live *"is not universally valid in its objective expressibility"*[8] but it is decisive for how we shape our lives, what values, what meaning we attach to them. It is concentrated *"in thought contexts that permeate a life as a whole"*. It is closely linked to the person who thinks it and, ideally, lives it. But since it is not universal, it may be that a philosophical text which attempts to express such a truth may open up a profound meaning to one, but seem utterly meaningless to another.

---

[8] Jaspers (1960), p. 115.

Therefore, philosophy can never arrive at definitive results like the natural sciences.

Jaspers presents Socrates as the ideal image of a philosopher: *"To have Socrates before one's eyes is one of the indispensable prerequisites of our philosophizing."*[9] Socrates himself left no writings, especially no elaborate argumentations on philosophical topics. *"In the context of the history of Greek philosophy as a history of theoretical positions, he has no place."*[10] Instead, he stands for constant questioning. *"I know that I know nothing"* is his most famous saying. It has been handed down that he constantly went about Athens tormenting the people he met with his questions, thus forcing them to think, and then subjecting their answers to rigorous criticism through follow-up and further questioning. He suspected that philosophy is always on the way and never comes to an end, that the essential thing is the conscious ever deeper penetration into the questions as well as the critical illumination of the always provisional answers.

When Socrates "maltreated" his fellow men, he always addressed only each individual for himself, because philosophy is an individual matter. Since it does not give universally valid answers, everyone must go through the process of questioning himself, must wrestle with the questions and pull out what is essential for him. Kant writes in the *Critique of Pure Reason that* one cannot learn philosophy, but only how to philosophize. Jaspers sees it the same way. Traditional philosophy (or the history of philosophy) with its recurring themes can be an inexhaustible source of inspiration that helps shape the directions of one's thinking, but it does not offer content that can be studied as universal truths like the content of the sciences.

Jaspers emphasizes that philosophy must deal with one's own life as well as with the findings of science:

> *A self-isolating philosophy would be without reason. Philosophy as a discipline remains a questionable thing. As a doctrine it only creates attention. The study of philosophy, then, occurs through the study of the sciences and through the practice of one's own life, awakened by the great philosophy of tradition.* (Jaspers 1960, p. 284)

This kind of philosophy can never reach an end. A philosophy that provides definite answers and therefore makes clearly defined progress has never existed. Philosophy as a personal quest and journey of discovery is new again for every

---

[9] Jaspers (1960), p. 437.
[10] Jaspers (1960), p. 418.

human being to some degree. It cannot end as long as we humans have not yet forgotten to marvel.

In this book I will conduct philosophy of physics. I will do this somewhat differently from the way it is done in academic works dealing with the philosophy of physics. There, it is mostly about certain *positions* that one can take on the individual theories of physics, e.g. quantum mechanics or general relativity, about the "ontology" of the respective theory (i.e. the question of what in it are the things that really actually exist), about the "identity" of particles, about the meaning of spacetime, and so on. I am more concerned here with an overall picture, the world view that the theories draw *together*. This picture is far from complete; I will therefore not arrive at a strictly defined *position,* but I can give you the overall impression I have gained in the years of my research in theoretical physics, and in the years of thinking about it, which continue to this day.

In doing so, I would like to express the tremendous wealth of thoughts and structures, of ways of looking at the world, that come to light in the various theories. In summary, I want to describe the general character of things and facts in the world of physics, as it becomes visible in the interplay of the theories. Finally, my point is that we must learn to acknowledge the *limits of* physics, both practical and fundamental. On the other hand, I would also like to contribute something to the salvation of philosophy.

3

Mathematics

At school we get the impression that mathematics is mainly about arithmetic, arithmetic seems to be its core, its real meaning and purpose. All the time we learn and practice arithmetic rules that we expect to be able to apply at some point in specific tasks that we encounter in "real" life, for example at work or in regulating our finances. If we can't see such a possible application, learning them feels like a waste of time.

The picture of mathematics that I would like to draw here looks quite different. In it, mathematics is a world of its own, in which there is an infinite amount to discover, things of great beauty, full of strange connections. It represents a value in itself, completely independent of the question of what of it can be applied to practical problems in the other, physical world. That our physical world is itself so steeped in mathematics, more and more so the deeper one looks into things, only makes the matter all the more splendid and mysterious. Thus, understanding the world of mathematics also ultimately serves the deeper understanding of our own world. So feel all the pure mathematicians in the universities and some outside them who devote their lives or part of their lives to the exploration of that mathematical world. In universities, pure mathematics is distinguished from applied mathematics; the former is concerned with structures for their own sake, the latter with areas that promise special utility for certain purposes outside mathematics.

Unfortunately, it is difficult to convey to an outsider what these people are actually working on and what magnificence they get to see in the process. This is because pure mathematics is not about objects we know from our everyday lives or to which we can directly relate. Worse, mathematics works with multiply nested cascades of definitions (as we will see in a bit more detail in a moment), and its study requires, above all, the patient step-by-step learning of these "vocabularies." Therefore, it is quite impossible to explain the work on certain structures in everyday language. The laudation of the winner of a mathematical award is therefore mostly a difficult undertaking. It is still easiest with work in the field of number theory, because the integers are familiar to us from everyday life. That is why the solution of Fermat's problem by Andrew Wiles could be popularized quite well. In fact, for many pure mathematicians, the fascination begins in childhood with number theory.

The popular scientific literature on mathematics has found various ways to cope with this "language problem". For one thing, there are at least some areas of mathematics that require somewhat less convoluted vocabulary and are somewhat more familiar to us from everyday life. These include number theory, Euclidean geometry, and probability. These topics can be written about quite well. Second, there are many eccentric or otherwise interesting characters in the history of mathematics whose lives can be well told. Third, some mathematicians have for some time been trying a new way of communicating their fascination with mathematics: They accept that their description or intimation of the facts may be incomprehensible to a large proportion of readers, but in doing so they describe their own involvement with the subject so vividly that the spark is nonetheless ignited. These books include *Birth of a Theorem* by Cedric Villani (2015) and *Love and Math* by Edward Frenkel (2013). In particular, I would recommend the latter book. Some of the mathematics presented in it is, as I said, difficult for an outsider to follow, but Frenkel so deftly weaves this into the telling of his own life story and conveys his love of mathematics so vividly that this book conveys what mathematics means to the initiate more compellingly and clearly than almost any other.

At this point I would like to bring up a somewhat vexed subject. Many people question the raison d'être of such "useless" disciplines as pure mathematics. The criticism is often expressed in dramatic terms: "How can we bother with such things? How can the state pour money into this kind of research while people are dying in the world? Why don't they rather invest the money in hospitals?" or something like that. Of course, the same question is asked not only about pure mathematics, but also about other "useless" areas of basic research, such as astronomy or particle physics. I must confess that this criticism is repugnant to me and I find it difficult to restrain my distaste when it is voiced.

I also dislike the way in which those responsible for science often respond to this: "It is just the case with basic research that its usefulness only becomes apparent much later, it cannot be foreseen beforehand. It is therefore good to pursue many directions, some of which will sooner or later yield unexpected benefits." First of all, that's a lie. For many research projects, it's pretty clear that practical benefits are pretty much out of the question. Moreover, this response embraces the imperative of utility, bows to it.

A much better answer was given by Michael Ende. He was not known to be a great friend of mathematics. His answer was directed against those who accused him that his literature was not political enough or did not have a clearly definable message. But it applies equally well to "useless science". Indeed,[1] in a letter to a reader, he writes: *"A good poem is not there to improve the world – it is itself a piece of improved world."* These "useless" things represent a value that has a meaningful effect. After all, we don't say, "If Beethoven had once worked in a hospital instead of writing useless symphonies." Or, "If only Goethe had spent more time on his building projects instead of producing beautiful literature." We don't ask ourselves after every visit to the cinema or when we treat ourselves to something else whether we shouldn't have donated the money instead. When confronted with the question, "Why should humanity actually be protected and saved from ruin, when it does so much harm, to each other and to nature, when it displays so much greed, envy, selfishness, complacency, anger, what is so worthy of protection about it?", then one of the best answers is: "Because it is capable of creating and feeling beauty, because it expresses a love for things and for each other that goes deep and wants to fathom, beyond purely pragmatic utility, this wanting to understand and research, the inspiration and creativity, the skill and mastery that it is capable of developing in the process, the works of lasting value that are created as a result, the connections that are made visible."

Of course, it depends on a suitable balance. We cannot all sit in the ivory tower and indulge in aesthetic; the number of problems and the amount of suffering on this planet are great. But let's look at the ratios: The total gross domestic product of all the countries on earth combined in 2018 is about $80 trillion (80,000 billion). About 2 trillion of that, about 2%, is spent on military purposes. Only a few billion, less than 0.01%, goes to such research projects that serve no purpose other than to deepen our understanding of the world. The value that this creates may not be immediately obvious to everyone. But it is substantial, and most importantly: meaningful. If we abandon beauty and depth and devote ourselves only to the "useful," our world will quickly look like a socialist prefab. Nor would we benefit from devoting all of our time and effort to just chasing after suffering and inconvenience in the world, because that's never going to be finished (though of course I'm all for

---

[1] Cited in https://de.wikipedia.org/wiki/Die_unendliche_Geschichte

devoting a *lot* of our time and effort to that; the issue for me is the *exclusivity* that is propagated by some). Different people feel called to different tasks. It's great that for many, that task is to help others live better lives. But it is equally great that some feel called to create something positive that exists independent of human (and animal) suffering, something that can be marveled at and found beautiful, and that can lead us to deeper insights about this world or even other worlds. These things include the arts, but they also include, precisely, pure mathematics, astronomy, and particle physics. Really, a good poem is itself a piece of an improved world.

3.1   Sets and Structures

The foundation of mathematics, in the modern view, is **set theory** (or, as some mathematicians would insist, its even more modern extension, **category theory;** we will content ourselves here with sets). What a set is, Georg Cantor characterized in a famous definition:

> *By a* **set** *we mean any aggregation of certain well-distinguished objects of our view or thought (which are called the* **elements of** *the set) into a whole.* (Cantor 1895)

So, certain objects, whether concrete or abstract, are combined into a whole, e.g. the set of hairs on your head or the set of prime numbers. The crucial thing is that the objects are all distinct, i.e., the same element cannot be in the set more than once. A set can be finite (such as the hairs on your head) or infinite (such as the set of prime numbers; there are infinitely many of them). A finite set can be characterized by properties of its elements or by enumerating its elements (with infinite sets, enumeration becomes difficult). For this, one uses the notation with curly brackets and commas. For example, the set of integers from 1 to 10 is the same as the set $\{1, 2, 3, 4, 5, 6, 7, 8, 9, 10\}$.

Mathematical structures are formed by placing several sets in certain relationships to each other or by placing the elements of a set in certain relationships to each other. The notion of a **function** plays a crucial role here. A function from a set M to a set N (notation: $M \mapsto N$) is a rule that assigns an element of N to every element of M. The names M and N are completely arbitrary. It is common in mathematics to use such short names or symbols for everything. We could also call the sets Gertrud and Rita, but that would be unwieldy. Some sets have fixed symbols by convention, such as the set $\mathbb{Z}$ of integers or the set $\mathbb{R}$ of real numbers. For example, we can define a "length function" $M \mapsto \mathbb{R}$ from the set M of hairs on your head to the real numbers that assigns each hair its length in centimeters. A function corresponds to a table with two columns, where the first column contains each element of the "initial set" M exactly once, and the second column contains an assigned value of the "target set" N.

One can combine sets by **Cartesian products.** The Cartesian product $M \times N$ of two sets M and N consists of all pairs (m, n), where m is an element of M and n is an element of N. For example, if $M = \{1, 2\}$ and is $N = \{3, 4\}$ then $M \times N = \{(1, 3), (1, 4), (2, 3), (2, 4)\}$. A plane can thus be viewed as the Cartesian product of two straight lines: A point x on o ne straight line and a point y on the other straight line are combined to form a point (x, y) of the plane.

An **operation** is a function $M \times M \mapsto M$, so two elements of M are assigned another element of M. In this way the arithmetical operations come into play. For each of the arithmetical operations "plus", "minus" and "times" is such a function on a suitable set of numbers, e.g. $\mathbb{Z}$ or $\mathbb{R}$. A pair (a, b) of numbers is thereby assigned the number $a + b$, $a - b$, and $a * b$ resp.

The above already shows how we have to deal with nested conceptualizations in mathematics. On the first level was the notion of a set, on the second the notions of a function and a Cartesian product, which build on the notion of a set. On the third level we find the notion of an operation, which builds on the notions of a function and a Cartesian product. In this way it continues. Level by level the edifice of mathematics is raised. Each level contains definitions that build on the previous levels. Each definition corresponds to a new vocabulary word that the student must learn. During a study of mathematics, thousands of such vocabulary words come together. In learning a normal language, the words can be learned independently of each other because each word has an equivalent in our own language that can be translated to the foreign word. In mathematics, however, each vocabulary word builds on lower level vocabulary words, and there can be a lot of levels. These "higher" terms also have no equivalents in our everyday language; they are just abstract terms tailored to specific mathematical structures. Therefore, one can only learn the vocabulary of mathematics in a certain order and not just jump directly to a higher level or "explain in simple words" what these levels are about. This is why it is so hard for a pure mathematician to explain to a non-mathematician what they are working on. An applied mathematician can at least still tell about the applications that benefit from his work.

Vocabulary often includes symbolic shorthand notations to be learned along with the terms, such as $M \times N$ for the Cartesian product or the curly braces for enumerating the elements of a set. Other examples are $a \in M$ for "a is an element of the set M" or $f(x)$ for "the element of N that the function f

assigns to an element x of M". Operations are special functions for which the notation with $f()$ is not common. We write simply $a + b$ rather than, say, $f_+((a, b))$ or $+((a, b))$. (The double parentheses are intentional. The outer brackets belong to the notation $f()$, the inner ones denote an element (a, b) of $M \times M$.)

To be able to indicate the further progress of things, we want to use here the notation $a . b$ for operations, as a generalization of $a + b$. That means, the dot can stand for any operation, for example for +, - or *. We continue with this notation at the fourth level. An operation is called **commutative** if $a . b = b . a$. Addition and multiplication are commutative, but subtraction is not, because $a - b$ is not the same as $b - a$. An operation is called **associative** if $(a . b) . c = a . (b . c)$. Again, addition and multiplication are associative, but subtraction is not, because $(3 - 2) - 1 = 0$, but $3 - (2 - 1) = 2$. A **neutral element** n has the property $a . n = a$ for all $a \in M$. For example, 0 is the neutral element for addition and subtraction, because $a + 0 = a$ and $a - 0 = a$. The number 1, on the other hand, is the neutral element for multiplication, because $a * 1 = a$. An **inverse element** $\bar{a}$ to a given element a has the property $a . \bar{a} = n$. For addition $-a$ is inverse to a, because $a + (-a) = 0$. For multiplication $\frac{1}{a}$ is inverse to a, because $a * \frac{1}{a} = 1$. The notion of an inverse element is already at the sixth level, because it presupposes the neutral element.

Finally, on the seventh level we encounter the **group**. A group is a set with an associative operation, which has a neutral element and also contains an inverse for each element. The set $\mathbb{Z}$ of integers is a group with respect to addition, but not with respect to multiplication, because $\frac{1}{a}$ is in general not an integer, so it is not contained in $\mathbb{Z}$.

Groups occur in a tremendous variety also in our physical world, they have numerous applications in physics and also in other sciences. This is mainly due to the fact that the successive execution of certain operations can be understood as an operation in the above sense. If you rotate any object twice in succession (in different directions, through different angles), the result is again a rotation, i.e., you could have reached the final state with a single rotation. If we denote the two rotations by d_1 and d_2 and the single rotation, with which we would have achieved the same result, by d_3, then we can formally express it like this: $d_1 . d_2 = d_3$. This time the dot stands for the successive execution of d_1 and d_2. The "neutral rotation" is the rotation which does not change anything, i.e. a rotation around the angle $0°$. For each rotation there is an inverse rotation, which undoes the first rotation. So all possible rotations together form a group. The same is true for permutations (interchanges) of a set of objects. If I rearrange the books on my shelf, i.e. put them in a new order, then I can do this several times in a row, and a group structure results, completely analogous to the rotations.

At level eight we learn about **subgroups**, which are subsets of a group that are also groups in their own right. In the case of rotations, for example, we can restrict ourselves to rotations about a certain axis. Combinations of such rotations are again rotations about the same axis, so they form a subgroup of the group of all rotations about any axis. At level nine we hear about the **cosets** that can be formed from a subgroup and the elements of the original group. At level ten, this leads us to the notion of a **normal divisor**, a special kind of subgroup that allows a group to be "decomposed" or "factorized" into two smaller groups. This brings us, on level eleven, to **simple groups**, which are groups that cannot be decomposed further, much like prime numbers in the set $\mathbb{N}$ of natural numbers.

A long-running project that became famous, started in the 1920s and essentially completed in the 1980s, is the **classification of finite simple groups**. About 100 mathematicians have worked on it, and their results run into tens of thousands of pages. A summary in "only" twelve volumes is currently in progress. Most of the infinitely many finite simple groups can be divided into 18 "families", that is 18 different construction rules, each of which can be used to construct an infinite series of finite simple groups. However, there are exactly 26 exceptions, the **sporadic groups**, which do not fit any of the 18 patterns. The largest of the sporadic groups is the so-called **monster group** with about $8*10^{53}$ elements; this is a number with 54 digits. Much of the mathematicians' work has been to prove that there could be no other sporadic groups besides these 26.

3.2 Proofs

Proving is, next to defining, the most important activity of the mathematician at all, and it is the more extensive of the two. If set theory is the foundation of mathematical structures built over defining, logic is the foundation of proving. Any structure built up via definitions has properties that go beyond what is defined, but follow logically from it. To tap into these properties, to prove them, is to understand the structure better.

Properties are formulated in the form of propositions, e.g. "There are infinitely many prime numbers". This statement presupposes that we have already sufficiently defined the structure of the natural numbers $\mathbb{N}$ and understood it to such an extent that we have developed from it a concept of divisibility and from it the concept of prime numbers (a number divisible only by 1 and itself). Now we are at the point of wanting to understand what logically follows from this notion. The statement "There are infinitely many prime numbers", until we have proved it but consider it a promising candidate for doing so, is a **conjecture**. By proof it becomes a **theorem**.

Very often words that are familiar to us from everyday life are given a different meaning in mathematics. A group in mathematics is something different from, say, a group of people. A "simple" group is in many cases anything but simple in the conventional sense. Rings and fields are also names for certain structures that have nothing to do with the meaning of these words that we know from everyday life. Because they have so much to define, mathematicians often make use of existing words to which they then give a new meaning, simply so as not to have to introduce too many fancy names. Who knows what mathematics would look like if it were done by people like Tolkien.

Back to proof. Logic gives us certain rules and techniques on how to draw correct conclusions and build proofs. One technique that is used very often is **indirect proof**, also called **proof by contradiction**. In this, one starts from the assumption that the statement one wants to prove is false, and leads this assumption to a contradiction. Because, according to logic, if the falsity of the statement is contradictory, then the statement must be true. This is also how one proves that there are infinitely many prime numbers. So let's assume the opposite of that, that there are only finitely many primes, namely $p_1, ..., p_n$. Then we can multiply them all up and get $N = p_1{}^*p_2{}^* ... {}^* p_n$. The number N is therefore divisible by all prime numbers. But then $N + 1$ is not divisible by any prime, because the next number after N *that* is divisible by a prime p_i is $N + p_i$. Thus, however, $N + 1$ would be a new prime, as opposed to the assumption that the list $p_1, ..., p_n$ was already complete. So the assumption that there are only finitely many prime numbers was wrong. It follows that there must be an infinite number, which was to be proved.

This is a very simple proof. There are more difficult ones that span hundreds of pages. There are also countless open problems, conjectures that have not yet been proven. Sometimes a conjecture is unexpectedly disproved. Or it happens that one can prove that a conjecture can neither be proved nor disproved. This then means that the underlying structure has not yet been sufficiently constrained by definitions to guarantee that the statement is true or false. The structure can then be further refined, in either way: The statement can become true by one particular refinement and false by another. The best known example of this is the development of non-Euclidean geometry.

The art of proof was already cultivated by the ancient Greeks. The most comprehensive and mature evidence of this is provided by the writings of **Euclid of** Alexandria from the third century BC. From him also comes the proof that there are an infinite number of prime numbers. Most importantly, Euclid left his mark on geometry. His exposition of the field was absolutely fundamental for two millennia. The mathematical knowledge, the systematic approach, the sophisticated methods of proof expressed in it are still astonishing when measured against the times. Euclid was also concerned to minimize the basic assumptions from which he could logically derive, i.e. prove, the entire theorems of geometry. He begins with some definitions for the basic building blocks of his geometry, e.g. points, straight lines, circles, angles. Then follow the five basic assumptions (the **postulates** or **axioms**) that are supposed to hold for these basic building blocks. From these postulates he deduces a large number of theorems of geometry, for example "The sum of angles in a triangle is 180°" or "If two angles in a triangle are equal, then two sides are also equal in length", and so on. On the other hand, he also showed how certain geometrical objects could be **constructed** with compasses and rulers. For more than 2000 years, when the compass and ruler were still the essential tools for making drawings, especially floor plans for building projects, these methods were used. With today's computer methods, construction with compasses and rulers is certainly not quite as significant any more. The theorems of geometry, however, are timeless.

The five postulates can be understood today as a definition for a certain mathematical structure, the **Euclidean plane**. At that time, they were considered to be statements about reality that no one with common sense can doubt and therefore do not require proof. Of the five postulates, four are very simple, for example, "Any point can be connected to any other point by a straight line." The fifth postulate, called the **axiom of parallels**, on the other hand, is a bit more "unwieldy". It states, in modern terms, that for every straight line g and every point P that is not on g, there exists exactly one straight line h that contains P and has no point of intersection with g. ("No intersection with g" is the same as "parallel to g" according to the Euclidean definition, hence the name parallel axiom). This fifth postulate was a thorn in the side of mathematicians. For 2000 years they tried to prove it from the other four postulates and thus to reduce the number of independent basic assumptions to four – without success, until in the 1820s some mathematicians finally had the idea to try it the other way round: They constructed a mathematical structure that satisfies the first four postulates but not the fifth. This was possible without contradiction.

Thus several things were gained. First, **non-Euclidean geometry** was born. Second, the independence of the fifth postulate from the other four was proven. Thirdly, it was shown that the fifth postulate could be doubted with common sense. That there is exactly one parallel to every straight line through a given point is a fact of experience, of which we can convince ourselves on a flat piece of paper, a table top, or a flat floor. But this does not prevent us from defining a surface curved in a certain way on which this is no longer true. Fourth, it was an important step toward the modern view that mathematical structures can be defined *independently of* the facts of experience. The numbers and geometrical constructions we know from everyday life are a beginning and a source of inspiration, but by definition mathematics can be pushed far beyond wherever our wealth of ideas, sheltered and hemmed in by the laws of logic, takes us intellectually. Fifth, the axiom of parallels is recognized as a distinguishing feature between different geometries. With the first four postulates, geometry is not yet sufficiently specified to decide whether the axiom of parallels holds. From here, one can now further specify the geometry in several ways, either by adding the parallel axiom, or by adding another postulate that contradicts the parallel axiom (for example: there are no parallels, or there are more than one, or sometimes this way, sometimes that way, depending on the point and straight line).

3.3   Mathematics: Man-Made or Man-Independent?

Historically, at the beginning of mathematics were arithmetic (calculating with numbers), geometry and logic. How can we imagine the process that led to all the abstract generalizations, the manifold structures that mathematics deals with today? We have just discussed one example, non-Euclidean geometry. The notion of a set and the associated set theory was founded by Georg Cantor at the end of the nineteenth century. But Dedekind, Peano and Frege also developed similar ideas in parallel. The time seemed ripe for it. Cantor's version prevailed only in the twentieth century and was further refined by Zermelo and Fraenkel until 1930. It was also around this time that the idea of set theory as the foundation of all mathematics took hold. The study of groups was set in motion by Evariste Galois around 1830, when he investigated the symmetry of the solutions of certain equations, called polynomial equations, before he died in a duel at the age of only 20. Several more decades passed, however, before the definition was expressed in its present form in 1882. In the intervening time, the term underwent several clarifications and generalizations. One develops a term and finds different examples expressing different aspects. One draws conclusions, tweaks the definition to make it clearer, to exclude ambiguities, or to expand it, to include a larger class of structures under the same term. Or one finds contradictions that need to be ironed out by improvements. In the end, clarity, interestingness, aesthetics, applicability, and sometimes simply the "mindset" of the mathematicians of the time determine which terms and definitions ultimately prevail.

Now this may sound as if a certain arbitrariness is inherent in mathematics. Detached from the world of experience, the mathematician can define whatever he wants. In reasoning (proving), he can take as a starting point whatever he wants, and from there think in a direction he chooses. How the overall edifice of mathematics develops seems to be solely up to the tastes of the mathematicians. But perhaps it is similar to hiking: The hiker decides from where to where he goes and what equipment he takes with him, and possibly follows certain fashions, e.g. to walk a certain pilgrimage path that is popular at the moment. But the landscape he walks through exists independently of him. Taste makes him prefer certain landscapes, but in doing so he is only choosing from something given; the character of each landscape exists even without him. The view from a certain point is so good because it is the top of a mountain. The hiker who is eager for a view therefore chooses it naturally.

It is an age-old philosophical question whether mathematics is *created* or *discovered* by man, that is, whether it exists independently of us and we merely wander about in it (to use the image above), or whether it is a mere mental construction whose character and existence depend on the tastes of mathematicians and which has no higher truth value. Both views are held and defended by many clever people. The question is difficult to decide, and the concept of existence in this connection very obscure. After all, how are we to know if mathematics "exists" independently of us (a very Platonic thought), and what does that even mean? Perhaps we should still distinguish between two things here. Once a structure is clearly defined, then propositions that hold for that structure, and proofs of those propositions, are *discovered*, not invented in addition. For by defining the structure, everything within it is already established. The question is rather whether the structures themselves are invented or discovered.

This is where the **uniqueness of** mathematics comes into play. The terms in mathematics are so clearly defined that true and false can be quite clearly distinguished from each other. This is quite different from everyday language. Let's take the statement "Otto is goofy." The criteria for goofiness are not clearly enough defined to declare this sentence unambiguously true or false. There is probably some evidence for Otto being goofy, and some evidence against it. In the end, it's a matter of judgment whether or not you agree with the statement. Even the propositions of the natural sciences are sometimes not as clear-cut as they seem. The statement "light consists of electromagnetic waves" is correct for most concerns, but there are also situations in which it is crucial that light consists of particles, the photons. In the natural sciences, a new discovery can always cause a known statement not to be revoked (or only in extreme cases), but nevertheless to be somewhat restricted in its validity, to point out exceptions. However, a mathematical sentence such as "There are an infinite number of prime numbers" cannot be shaken.

Mathematics forms a language unto itself and has its own grammar. An extraterrestrial civilization may have completely different means of expressing itself than we do. But it will still count "1, 2, 3 ...", even if it uses completely different symbols for it. That is to say, there is a *structure of* counting that can only be this way and no other, no matter what words and symbols and what form of linguistic expression a civilization uses. That is, 1 + 2 = 3 also does not depend on the linguistic expression of the civilization, it is a mathematically correct statement that can only be so and not otherwise, no matter in what syntax the civilization formulates or symbolizes it.

But if in mathematics both the definitions and the propositions which follow from them are so unambiguous and unmistakable, so independent of the linguistic background from which they are formulated, then the question remains as to *what selection* out of the infinite number of possible definitions a civilization makes, what structures (apart from the natural numbers, which certainly no civilization can get past) it considers interesting to study. But where its choices overlap with ours, it must arrive at the same results as we do, even though it may express them linguistically quite differently. At most, it may be that if A and B are logically related, we find A to be more fundamental and take B to be consequential, while they find B to be more fundamental and deduce A from it.

Even the choice is not completely arbitrary. Starting from problems of arithmetic and geometry, which arise in everyday life, both with us and – at least in a similar form – with an intelligent extraterrestrial civilization, in many cases one thing leads to another. New structures impose themselves to solve certain problems, or arise as extensions or generalizations of old ones. Connections become recognizable, which want to be fathomed further and require further conceptualizations. Also the laws of nature, at least the physical ones, are the same for the other civilization, and these are themselves mathematical in nature and correspond to quite specific structures. So there is something to be said for the fact that they, at least in large parts, make a similar choice as we do.

All this leads us back to the picture with the landscape, which is formed by the infinitely many conceivable mathematical structures, which exists independently of us and on which we only tread a certain path, on which we select certain structures, triggered by problems from everyday life or the natural sciences, but continued in pure mathematics from the pleasure of wandering, motivated only by the hope of beautiful views and the fathoming of deep connections. The "correctness" of this picture can certainly not be proven by logic or by scientific methods. This is not even necessary, because it is only a picture. It is enough that it seems coherent to us and corresponds to the experience of a large number of mathematicians who have the impression that they are discovering and not inventing.

The topic of aesthetics, of beauty in mathematics, has already been briefly touched upon. Certain structures or even proofs are considered "beautiful" by mathematicians. For many laymen, mathematics is exactly the opposite of beautiful, only they tormented themselves with it in school and are glad when it is finally over. It's much the same with the taste of wine. When we taste it for the first time as children, we grimace in disgust. Even more so, we can't tell a fine wine from a cheap one. The taste must first develop, slowly mature, be cultivated. But some people never acquire a taste for it. It's the same with mathematics. You first have to delve into it in order to slowly understand what it's all about; the harmonies and connections that exist between the structures have to develop step by step in our minds, unfold their special aroma (yes, don't laugh!), then the sense of beauty already develops. It cannot be explained.

We can now summarize the characterization of mathematics: **Mathematics can be conceived as a kind of zoology of mathematical structures**. That is, mathematics deals with structures; these are sets in which, via various cascades of conceptualizations, certain relations are established by definition, via which propositions can be proved. In the process – as in zoology – classifications and sub-classifications of structures arise, most impressively seen in the classification of finite simple groups. Because there are infinitely many possible definitions, which thus lead to infinitely many mathematical structures, and because an infinite number of propositions can be proved for each of many structures, mathematics is **infinite**. Its definitions, theorems, and proofs are **exact** and **unambiguous**. It is also **timeless**: its structures are not subject to change (as in the picture with the landscape), only how we deal with the structures, which names and which notation we use for them, which ones we regard as important, in which order we examine them and build on each other, that is subject to the changes of time. It is perceived as **beautiful** by mathematicians, the aesthetic aspect plays a major role.

3.4   Numbers and Spaces

Since this book is primarily about physics, we will now take a look at which mathematical structures are particularly important for physics.

With the progress of arithmetic, the concept of a **number** became broader and broader. In the beginning there are the **natural numbers** $\mathbb{N} = \{1, 2, 3, \ldots\}$. With the addition of the zero and the negative **numbers** $\mathbb{Z}$, the **integers** are created. In order to complete the set with regard to division, this set is extended by the fractions. Thus one arrives at the **rational numbers** $\mathbb{Q}$. It can be shown, however, that certain numbers which occur in geometrical contexts or as solutions of certain equations cannot be written as fractions, e.g. $\sqrt{2}$ (the length of the diagonal of a square with side length 1) or π (the ratio of the circumference of a circle to its diameter). This brings us to an even larger set of numbers, the **real numbers** $\mathbb{R}$. The functions that are treated in school mathematics are mostly real functions, i.e., the initial set is $\mathbb{R}$ or a part of $\mathbb{R}$, the target set as well. Physical quantities are almost always understood as real numbers, supplemented by a unit of measurement such as centimeters, seconds, or degrees Celsius. However, the real numbers can also be extended to include the roots of negative numbers. As is well known, these are not contained in the real numbers, which is why you learned the rule in school that you are "not allowed" to draw a root from negative numbers. With the **complex numbers** $\mathbb{C}$, these roots are introduced simply by definition, by setting an "imaginary unit" i as $\sqrt{-1}$. The complex numbers have many nice properties. In fact, the behavior of many functions can be better understood by extending them from the real to the complex numbers. Equations can be solved better, and certain structures (such as the so-called Lie groups) can be classified more easily. Although the construction with the imaginary unit seems somewhat artificial at first, after some practice and understanding, mathematicians find the complex numbers more "natural" than the real ones. In physics, the complex numbers become indispensable only in quantum mechanics. In this theory, there are states which, simply by nature, can only be expressed in complex numbers.

With each extension of the number range, the operations that can be performed and/or the solvability of equations change. For example, division is possible in $\mathbb{Z}$ only in certain cases (divisibility), in $\mathbb{Q}$ always, unless you try to divide by zero. For equations with one or more **unknowns,** we look for **solution sets** whose elements satisfy the equations. The equation $x + 1 = 2$ has as its only solution $x = 1$, so the solution set consists of a single number (whether we look for the solution in $\mathbb{N}, \mathbb{Z}, \mathbb{Q}, \mathbb{R}$ or $\mathbb{C}$). In the equation $x - y = 0$ with two unknowns, we look for pairs (x, y) that satisfy this equation. There are infinitely many solutions, namely all pairs where x and y have the same value, $x = y$. The solution set consists of all pairs with this property. However, the equation $x^2 = 2$ has no solution in $\mathbb{N}$, $\mathbb{Z}$ or $\mathbb{Q}$, but two solutions, namely $\sqrt{2}$ and $-\sqrt{2}$ in $\mathbb{R}$ or $\mathbb{C}$. The equation $x^2 = -1$ has no solution in $\mathbb{N}$, $\mathbb{Z}$, $\mathbb{Q}$ or $\mathbb{R}$, but two solutions, namely i and $-i$, in $\mathbb{C}$.

As we saw at the beginning of the chapter, the arithmetical operations that we know from the various number systems have properties that can be further generalized and abstracted, which has led to the general concept of an operation. The field of mathematics that deals with these operations is **algebra**. In algebra, one studies structures in which certain properties are ascribed to the operations; as an example, we have discussed groups. These have numerous applications in physics, especially when it comes to **symmetries of** physical systems.

From the side of geometry, generalized notions of **spaces** have developed. (The demarcation is not so clear, by the way; geometry and algebra overlap in many aspects, for example the real numbers themselves form a one-dimensional space, and the complex numbers can be conceived as a two-dimensional plane). One can take the notion of space so broadly that it applies to virtually any mathematical structure. Somewhat more restrictedly, one can say that a space is a **continuum of** points, i.e., a set whose elements we call points, and which are so **close** together that in every neighborhood of a point, no matter how small, other points can be found. This is not a sufficient mathematical definition, because we would first have to define what we mean by *neighborhood*. But it gives a first, halfway descriptive picture of the concept of space. We don't want to rattle off another cascade of definitions here. There are completely different kinds of spaces in mathematics, they are called vector spaces, affine spaces, Banach spaces, Hilbert spaces, etc., and they all differ in what structural elements are defined on them and what properties are set for them.

Typical terms defined on spaces are lengths, distances, and angles, as well as topological terms that characterize the global properties of spaces. On a **vector space**, the points can be added together; they are called **vectors** there. In physics there are many quantities that are described by vectors, for example forces. Forces of different magnitude and direction acting on the same object are added together to form a total force.

A common characteristic of spaces is their **dimension**. The dimension of a space is a natural number that describes the number of its independent "directions", or the number of coordinates needed to define a point. A straight line is one-dimensional, a plane is two-dimensional, and the "normal" space we live in is three-dimensional. However, one can easily define spaces with arbitrarily high dimensions, especially infinite-dimensional spaces. For the layman, it is then often said, "You can't imagine such a thing." But the mathematician has a different conception of imagination; for him the higher dimensions pose no problem. His imagination is accustomed to the abstract concepts and is geared to them.

By the way, it is similar with large numbers. As a layman, you want to "imagine" a number in a certain way, for example, by picturing certain objects in that number. Beyond a certain size, you say, "I can't imagine that." When the news mentions the size of an area, e.g., a forest fire or oil spill, typically another comparison is made, such as "That's an area as big as the Saarland," so that we can "imagine" something under the number. Similarly with size ratios, e.g. when an educational lecture mentions the size of the atomic nucleus compared to the entire atom: "That's as big as a grain of rice in a football stadium." That way the layman can "visualize" it better. The billions of light-years that comprise the part of the universe visible to us are considered "unimaginably large distances" from the outset. To a mathematician, this seems rather amusing. For him, the idea of a sextillion (10^{21}) represents no greater hurdle than that of the number 5. His imagination works with a different kind of images that have developed from the constant handling of abstract concepts.

In physics, all kinds of spaces occur. Vector spaces we had already mentioned. The three-dimensional space that surrounds us appears to us essentially as an affine space. With general relativity, however, we must conceive of it as the hypersurface of a four-dimensional curved Riemannian space. Hilbert spaces play a central role in quantum mechanics. We have not defined any of these terms here. The enumeration is only meant to indicate to you that many of the various abstractions that were tried out on spaces for purely mathematical interest suddenly took on unexpected meaning in physical reality.

3.5 Differential Calculus

The decisive factor for the birth of physics in modern times was the **differential calculus**. It was developed independently by Isaac Newton and Gottfried Wilhelm Leibniz in the second half of the seventeenth century, which led to fierce disputes between the two, which Newton in particular fought out with all ferocity and nasty tricks.

Differential calculus is first about **momentary changes in** physical quantities. Let's take **speed, for** example, that is, the change in a position in relation to the time it took to reach it. If you travel 100 km in one hour, then you have an average speed of 100 km/h in that hour. If you travel 50 km in half an hour, then you also have an average speed of 100 km/h in this half hour. In general, however, this speed will not be constant; you will have to brake and accelerate again and again in between. The speedometer shows you your current speed to a certain approximation. It measures how many centimetres you have moved in a certain fraction of a second, and to make it easier for you to understand, it then extrapolates the result back to an hour (this would then be the distance you would move in an hour if you maintained your current speed exactly). This measurement only approximates your current speed, because even in that split second, your speed will not be perfectly constant. There will be small fluctuations even in that short period, caused for example by bumps in the road. You would get the exact instantaneous speed if you made the time span over which the distance covered is measured **infinitesimal**, i.e. infinitely small. Distance and time then both go towards zero, but their ratio goes towards a constant positive value, i.e. the instantaneous speed.

In terms of measurement, we can only approximate the infinitely short time spans. But in mathematics we can calculate exactly with it and define terms (e.g. the instantaneous velocity), just with the differential calculus. The momentary change of a variable is also called its **time derivative**. The time derivative of the position is therefore the velocity. The "second time derivative" of the position is the time derivative of the time derivative, i.e. the momentary change in velocity, i.e. the **acceleration**. (Braking is considered to be a negative acceleration).

This can be generalized. **Spatial derivatives** are obtained by looking at functions not of time, but of location. For example, consider temperature as a function of height above sea level. You know that the air gets colder as you go up. Let's say it's 20 °C warm in a place by the sea. You go up in a balloon and measure how the temperature changes. At an altitude of 2000 m, you are at the freezing point, 0 °C. That is, on average, the temperature has dropped by 10° per kilometer. But the drop won't be completely uniform. Air circulation creates layers of air where it's a little bit cooler or a little bit warmer. The problem is completely analogous to velocity, you just need to replace "instantaneous" with "local". You get the local rate of change of temperature at a given altitude by looking at the temperature change on an *infinitesimal* increase of altitude. Again, this can only be approximated by measurement, but you can do wonderful calculations with such a spatial derivative.

Since physical space is three-dimensional, there are three independent spatial derivatives. You can look at the temperature change when you go up, but also when you go east or south.

Physical quantities are usually considered to be either functions of time, for example the functions $x(t)$, $y(t)$ and $z(t)$, which describe the three coordinates

(x, y, z) of a moving object as a function of time t, or as a function of both time and location, for example the temperature $T(x, y, z, t)$.

Equations that relate the various derivatives of functions and the functions themselves are called **differential equations**. Almost all laws of nature in physics are given in the form of differential equations. The wave equation, for example, which describes all wave phenomena, relates the second time derivative of a quantity to its second spatial derivatives.

The *solution set of* a differential equation consists not of numbers but of functions. One wants to know: If I know something definite about the derivative of a function (this information is given by the differential equation), what can I say about the function itself? Let's look at this with a simple example. You are moving at a constant velocity v in the *x-direction* (this makes things particularly simple, the instantaneous and average velocities are the same because the velocity is constant). What is wanted is your position $x(t)$ as a function of time. The differential equation is $\dot{x}(t) = v$ (here the dot stands for the time derivative). So you know something about the derivative of the function and now you want to infer the function itself from it. The solution set consists of the functions $x(t) = x_0 + v^*t$ with any real number x_0. The constant velocity means that your position changes proportional to time. This is expressed by the term v^*t. But the differential equation says nothing about the point at which you *started* the motion. Therefore, the solution set is infinite: one solution for every possible starting point. This starting point represents an **initial condition** that you must consider when choosing a solution from the solution set. It is determined by the parameter x_0 that expresses your position at the time $t = 0$. (In physics, clocks are usually thought of as stopwatches that are set to zero at the beginning of a process).

The parameters in the solution sets that are not fixed by the differential equation itself (x_0 in our example) are generally called **integration constants**. They always occur, in all differential equations. They are not always identical with the parameters describing the initial conditions of a process, but there is always a clear connection between the initial conditions and the integration constants.

Therefore, a physical process is usually described by two things: the differential equations that represent the laws of nature according to which the process takes place, and the specification of initial conditions (or integration constants) that pick a particular solution out of the set of solutions to the differential equations.

4

Natural Science

Natural science is *the* success project of modern times – the basis of all our technical achievements. In contrast to metaphysical speculation, it enables us to make *objective*, verifiable statements about the world. How does it manage to do this? Let us begin our investigation of this question with a definition that may seem somewhat unusual.

Natural science is a certain way to reach **consensus**. The content of this consensus is then called "knowledge" about "nature". The quotation marks are meant to express that knowledge and nature are terms to be used with caution, the meaning of which we will only gradually discover in the course of the discussion.

This definition seems unusual because we generally imagine that natural science is first and foremost about "truth," not consensus. But as a dutiful philosopher, one always has to be a little careful before bringing truth into play; so we start with consensus first.

Now, what is this particular way of reaching consensus that we call natural science? Many philosophers of science have tried to answer this question, and we must confess that it has not been possible to reach complete consensus on it. This is partly because the question of what is natural science is not itself a scientific question, but a philosophical one. Nevertheless, there are certain characteristics of natural science which can be emphasized without encountering too much contradiction, and which summarize the essence of this field of human activity somewhat aptly:

© Springer-Verlag GmbH Germany, part of Springer Nature 2022
J.-M. Schwindt, *Universe Without Things*,
https://doi.org/10.1007/978-3-662-65426-2_4

1. Natural science is **empirical**. That is, unlike in mathematics, where the formation of concepts and the insights that result from them are tied solely to the logic and ingenuity of mathematicians, natural science refers to the "world" that presents itself to our senses, to the *phenomena* that we observe there. One task of natural science is to **classify** these phenomena. Another task is then to hypothesize relationships between certain classes of phenomena (or relationships within a class of phenomena). Many of these correlations are **causal** in character, that is, they are of the form: "If this phenomenon occurs, then that phenomenon results." We discuss causality in more detail below.

   In order for the hypotheses themselves to have an empirical character, i.e. not to remain pure speculation, they must be **testable**. The philosopher Karl Popper characterized this testability with the criterion of **falsifiability**: For a hypothesis to be accepted as scientific, it must be shown how it could be shown to be *false*, i.e., a group of conceivable observations must be described that would disprove the hypothesis. In other words, a scientific hypothesis must be open to empirical *attack*. The more such attacks a hypothesis is subjected to and survives – by, in the case of a deliberate experiment that *might* lead to an observation that disproves the hypothesis, instead making an observation that confirms it – the more credit the hypothesis will gain in the eyes of natural scientists.

2. Natural science involves **experimentation** where this is possible. In astronomy, you can only observe distant stars, but not influence the conditions that prevail on them. With smaller, more manageable objects, such as a test tube filled with a liquid, you can set certain conditions yourself, in order to thereby ask "nature" a certain "question": You can close the tube and thereby prevent the exchange of matter with the environment. You can heat it or apply pressure to it with a piston, and observe what the effect of each is. There are many types of experiments, but they all aim to allow certain classes of phenomena to be studied more closely than would be possible "in the wild". A typical experiment involves the following:

   (a) Experimental set-up intended to produce the phenomenon under investigation

   (b) Exclusion of interfering influences that would prevent the phenomenon under investigation or mix it with other phenomena.

   (c) Selectively changing a particular variable and observing how this affects the phenomenon under study.

(d) the use of measuring instruments to detect these effects more precisely than would be possible with our natural senses alone

(e) A protocol that summarizes the experimental setup and results in a way that other scientists can repeat the experiment and test the results.

3. Repeatability and testability require **objectivity**. Objectivity is a kind of elementary "that which can be agreed upon". It means that only those aspects of observations and experiments are used for analysis that can be readily transferred from one researcher to another. This generally includes quantitative statements rather than qualitative ones. Reading off the numerical value indicated by a measuring device is unambiguous and the same for everyone standing in front of the same device. Pure adjectives such as *light* or *dark, large* or *small* are relative and imprecise; there are always grey areas that are perceived as light by one person and as dark by another.

An example: The infinite variety of colours is divided into colour terms such as *red, blue,* etc. by different cultures, sometimes in different ways. For example, in many ancient languages there was no word for *blue*. What we now understand as blue was then perceived as a shade of *green*. Therefore, in an observation in which colors play a role, it is much *more objective to* use a spectrograph, which breaks down the light into the wavelengths that occur in it and records the proportion of each wavelength, thus again making a quantitative statement, instead of the color terms, which one derives from one's own visual perception. This, of course, already presupposes scientific knowledge that has led to the interpretation of light as an electromagnetic wave and to the construction of spectrographs. Thus, on the one hand, objectivity is a prerequisite for natural science, and on the other hand, the findings of natural science are in turn a prerequisite for further increasing objectivity.

In complete contrast to objectivity are value judgments such as *good* or *bad, beautiful* or *ugly,* or the personal feelings of the researcher. These have no place in scientific reports (although they certainly play a major role in the process of scientific work and the decisions made therein; in physics in particular, certain equations are perceived as *beautiful* and others as *ugly,* and this influences many physicists as to which theory they are willing to devote more time to; we will come back to this).

Psychology is a particularly difficult science as far as the criterion of objectivity is concerned, because here it is precisely about human feelings, and these cannot be measured in the same objective sense as a physical quantity. Research here has virtually no choice but to rely on the behaviour or on the verbal or otherwise articulated assessments (thus ultimately also behaviour again) of its human test subjects. For example, subjects are asked how happy they consider themselves on a scale of 1–10. The numerical value expressed on this scale is understood as the "measured value". It can then be examined which variables in the lives of the test persons are correlated with these measured values. It remains unclear to what extent this is about the actual feelings of the test persons. In fact, only objective facts can be investigated, such as the correlation between life circumstances and a certain *behaviour*, e.g. the expression of a certain numerical value in response to a certain question addressed to the test person. What inner processes in the subject produced the utterance of this particular numerical value is not amenable to objective analysis in this context. (It is of course conceivable that brain research has something to contribute here; instead of the behaviour of the person, the behaviour of the brain or its components is examined there. But to what extent this can be about the "actual" feelings is a difficult question, which we want to address later in this book).

4. Consensus is delayed as long as possible. This principle is similar to the presumption of innocence in criminal law: as long as nothing is proven, nothing is proven. Often enough, the data situation allows for several different interpretations. In this case, controversial views are explicitly desired and helpful for the progress of natural science. Each view should then be supported by its proponents as thoroughly and convincingly as possible, similar to a court case, where the role of judge is played by the entire community of natural scientists. This continues until new observations are made that exclude some of the alternative views, until finally only one remains, and consensus is thereby eventually enforced. Since consensus is always the goal in the end, such observations are to be brought about by purposeful "decision experiments".

The absolute classic among such controversies is the question of whether light consists of waves or particles – a dispute that was fought out over centuries and supported by all kinds of arguments from both sides until it was finally decided in the nineteenth century in favor of waves, at least for the time being. The punchline of this story is that in the twentieth century it turned out that light is made up of so-called quanta, which have both wave and particle properties, and so ultimately answered the question with a fair "draw".

This punch line also shows that the account above is very idealistic. Indeed, several difficulties arise on the way to ultimate consensus:

(a) We humans often tend to make premature judgments. This is quite natural, because in our lives we are often forced to make quick decisions without being able to see the whole situation and all the consequences of our actions; therefore, we are virtually trained and specialized in quick and thus premature judgments. What is often necessary for survival in everyday life is a hindrance in science, but even natural scientists are of course not free of such tendencies.

(b) Also hindering is the emergence of a "mainstream" that excludes or belittles dissenting views. This can happen for a variety of psychological and sociological reasons, not least to survive the fierce competition of building a scientific career, where it can certainly be an advantage to be of one mind with the majority. These effects have been described in detail by Lee Smolin (2006) in his book *The Trouble with Physics*.

(c) It is often not clear when exactly a hypothesis can be considered disproved. Often a hypothesis can be salvaged by making additional assumptions, and sometimes this makes perfect sense. Smolin gives the following example: If you see a red swan, you will not abandon the hypothesis that all swans are white, but you will look for the person who painted the swan. As we will discuss in detail in Chap. 6, many different assumptions always go into the interpretation of an experiment. Therefore, in very many cases there is the possibility of arguing that the experiment does not disprove the hypothesis it purports to disprove, but that one of the assumptions made in the interpretation of the experiment was wrong.

(d) The punch line in the answer to the question of the nature of light reveals another difficulty: the conceptual system with which we approach a problem may turn out to be wrong. In the terminology of nineteenth century Classical Physics, wave and particle are two mutually exclusive characteristics, and thus in this conceptual system the question of whether light is made of waves or particles is a genuine either-or question, admitting only one answer. In quantum physics, however, it is shown that both characteristics can be united in the same object, and in a way that would never have occurred to any physicist of the nineteenth century.

The next three characteristics of natural science should serve to overcome this difficulty.

5. Natural scientists form an **ethical community**. This is the term Smolin (2006) uses, and I find it very apt. Natural scientists must be able to rely

on each other. They must trust each other to present their findings truthfully, to the best of their knowledge. They follow a common goal, the "knowledge" of "nature", and this goal must take precedence over purely egoistic goals, which can also be achieved via falsified results and personal cronyism. They must also be willing to admit their own errors and mistakes when such have been demonstrated by others, or when one has found them oneself. Furthermore, a scientist must make the effort to follow up and critically review the work of colleagues. This constant checking of each other's work is called *peer review. The aim* here is to be as unbiased as possible, i.e. once again to put personal interests and preferences aside.

None of this is easy, and as is so often the case, such an ethical ideal is never quite achieved. Examples of falsification, inadequate verification, prejudice, personal cronyism are sufficiently well known. But it is almost existential for natural science not to stray too far from the ethical ideal. This ideal cannot be forced. But those who practice natural science know that it would be their downfall, that they could no longer achieve "knowledge" if the deficiencies became too great.

6. Natural science is open to **surprises**. Its findings always have a character of provisionality, of approximation, based on previous experiments and observations, but always in anticipation of future, new or more precise experiments and observations or new considerations that put what has been found so far in a different context. In the process, unexpected things can come to light that no one expected and that possibly go beyond the conceptual framework of previous theories.

7. An important guide in the jungle of hypotheses is the principle of Ockham's **razor**, which goes back to the philosopher William of Ockham (ca. 1287–1347). This principle states that simple explanations of a phenomenon are preferable to complicated ones. In particular, explanations that get along with fewer assumptions or fewer variables are to be preferred. Anything unnecessarily complicated should be cut off as if with a razor.

Basically, it is irresponsible to use the term "explanation" so naturally at this point. We recall the sentence by Wittgenstein[1]: *"The whole modern conception of the world is founded on the illusion that the so-called laws of nature are the explanations of natural phenomena"* In fact, we haven't yet dealt with the question to what extent one can speak of explanations at all in natural science. However, Ockham's principle is usually characterized

---

[1] Wittgenstein (1963), p. 110.

in terms of explanations. In order to stay in safe waters for the time being, we will therefore at this point simply understand an explanation as the construction of a chain of logical or causal connections, at the end of which stands the phenomenon to be explained. In this sense, then, "(1) all men are mortal; (2) Socrates is a man" would be an "explanation" of why Socrates is mortal. This does not quite have the depth we would expect from a "genuine" explanation, and it is precisely for this reason that we will need to address the nature of such "genuine" explanations later. In the context of Occam's razor, however, the simple version will suffice for us.

The application of the principle prevents a hypothesis that has not succeeded from being maintained by ever additional hypotheses. For example, according to the Ptolemaic view of the world, which was not questioned throughout the Middle Ages, all celestial bodies were supposed to move exclusively in circular orbits, at a constant speed and with the earth as the fixed centre of the world. Since this was not consistent with astronomical observations – the planets performed much more complicated motions than simple circular orbits as seen from Earth – additional assumptions had to be made to maintain the hypothesis. So-called *epicycles* were introduced, orbits that consisted of multiple circular motions intricately nested within each other. Copernicus and Kepler found a much simpler explanation: the Earth is not the center of the universe, but is itself a planet moving in an elliptical orbit around the Sun like the other planets. Epicycles were thus cut out of the edifice of natural science with Occam's razor.

The principle has proven itself useful many times in the history of natural science. If you think about why it actually applies, you come across a rather complex background. In part, it can be justified by probabilities: A more complicated explanation of a phenomenon often requires a larger combination of circumstances that is less likely than the circumstances by which the simpler explanation reaches its goal. There are also general heuristic arguments for the principle, designed to show why Occam's razor is generally the most efficient method of approaching "truth." The philosopher of science Ernst Mach, on the other hand, argues that natural science is a priori oriented toward an *economy of thought,* that is, *"recognizes the most economical, simplest conceptual expression as its goal"*[2] so that Occam's razor thus embodies a basic character of natural science that needs no further justification.

---

[2] Mach (1882).

8. The manifold hypotheses that establish connections between classes of phenomena do not stand separately, but are joined together to form larger contexts, a process at the end of which are **theories**. A theory is a logical structure that, from a few concepts and basic assumptions, organizes a large set of hypotheses into an overall context by deriving them from those basic assumptions through a chain of logical inferences. Basically, this is something similar to Occam's razor: a large number of hypotheses are simplified by recognizing them as the consequence of a small number of basic assumptions.

   For example, the representation of the planetary orbits as ellipses with the sun as reference point is – in the sense of Occam's razor – a better hypothesis than the epicycles with the earth as reference point. But this alone would not be called a theory. It was Newton who, with his theory of gravitation, traced both the elliptical orbits of the planets and the parabolic falling motions of objects on Earth back to a single equation, a single *law of nature*.

   In biology, the theory of evolution describes how the diversity of species developed. It does not "explain" the occurrence of each individual species in particular, because highly complex processes play a role in this, which include an element that could be characterized, roughly speaking, as "chance" (again, a difficult term to which we will return). But it does provide a set of basic assumptions that describe the mechanism by which evolution occurs, and that make it *plausible* in very many cases why a species changes in a certain way under certain environmental conditions, *adapts to* the environment in a certain way.

   It is the theories that give us the feeling to "understand" something with the help of natural science (again such a difficult word). We recognize logical and causal connections, arrange them into a larger context, which we take as an "explanation".

9. Natural science strives for **unity**. All the theories together are supposed to form a consistent overall building, in which *every* phenomenon has its place and which represents a complete image of the "world". The theories thereby stand in certain relations to each other, form a hierarchy (which is in no way meant to be judgmental), which we want to examine more closely in Chap. 5.

10. **Causality** takes precedence over **finality**. As mentioned above, many hypotheses in natural science describe **causal** relationships, that is, how one particular phenomenon (a process, an event) results in another. One says the second phenomenon occurs *because* the first one was there. What about the reverse, **final** case? Is it also said that something happens *so that*

something else will happen? Isn't that completely symmetrical to the causal case? In behavioral science and related disciplines, we do indeed talk about this: a squirrel gathers nuts *so that* it will survive the winter. But in inanimate nature, it seems inappropriate to talk that way. We do not say that the gas in space agglomerates *so that* it can form a star, but *because* it is forced to do so by gravity. Even the final (i.e. purposeful) behavior of living beings is ultimately attributed to causal relationships: Animals behave purposefully *because* the processes of evolution (natural selection) favor the emergence of such behavior. Causality is clearly superior to finality in natural science as a whole. Why is this actually so?

The answer is not so simple. On the one hand, this is due to an asymmetry of time, i.e. the differences between the past and the future. However, this asymmetry, as it turns out, is not as fundamental as it seems at first glance, but is rather related to very specific conditions that prevail in our universe. This will occupy us in the further course of the book. On the other hand, it is also due to the fact that natural science tries to keep the operation of a "higher will", which works towards certain goals, out of its hypotheses, and so far with success. However, this is also not something that is clear from the outset. In summary, then, we have to say that the preference for causality is a characteristic of natural science that stands on somewhat shakier legs than the others mentioned.

With these points, in my view, the fundamental traits of natural science are presented. The correlations and theories that are the result of this process represent a "knowledge", provided that they have been successfully tested in experiment and observation and the community of natural scientists has reached consensus on them. This kind of "knowledge" is about halfway between faith and the knowledge of mathematics. It is clearly stronger than a belief because it has been tested by empirical procedures and has survived attempts at falsification as well as the critical *peer review process* of natural science. Unlike faith, it is also objective in the sense that the observations and experiments that support this knowledge can be repeated by anyone who is able to muster the necessary resources to do so. This is true, with certain limitations, in the case of those observations that depend on the occurrence of certain external events over which the scientist has no control, e.g., a supernova (stellar explosion) in astronomy; in that case, the observer must still bring the time to wait for such an event. However, it is clearly weaker than the knowledge of mathematics, because it is always valid only within certain limits of accuracy, always represents only an approximation (more about this in

Chap. 6), and there is always the principal possibility that a future surprise invalidates it.

This knowledge brings with it a variety of benefits. First of all, it serves to *enlighten,* to debunk superstitions and misconceptions. Second, in many cases it enables us to make *predictions.* In some cases, such as planetary orbits, these predictions are very accurate. In other cases, such as weather prediction or long-term climate trends, they are somewhat inaccurate. In still other cases, such as the course of a disease, only statistical statements are possible. Thirdly, we can derive practical *applications* from it, i.e. develop all kinds of machines, devices and processes, medicines and weapons, new materials and new forms of communication. We can use it to save lives and destroy lives, or even to transform lives. The possibilities are almost limitless. Fourth, and this is the point of most interest to us in this book, we seem to be able to "understand" the "world" through this knowledge; it seems to "explain" to us the phenomena we observe. But is this even true? And what does it actually mean?

4.1   Our Common World

Let us first ask ourselves what knowledge is actually about. The world, as I wrote in Chap. 2, is first of all a construct of my mind. But the process of natural science, which aims at consensus and objectivity, presupposes some additional properties of the world in order for the whole thing to work at all: The world is not just a construct of *my* mind, but it is a *common* world that I share with others. For it is only through the medium of the *common* world that natural scientists can communicate with each other. Only through the *common* world can they observe the same phenomena and reach the same conclusions about them. That we *assume* such a common world is thus a basic prerequisite for the success of natural science. But this is by no means to say that this common world is "reality".

Forgive me for approaching the matter with such philosophical caution. Of course, we all obviously experience the world as a common world, and only very few doubt its reality. But perhaps you have also experienced, at least for a moment, something like a *solipsistic doubt,* a feeling that perhaps it is all just a kind of dream after all and that neither the world nor other people are real, as in the film *Abre los ojos* or its American remake *Vanilla Sky.* Apart from this doubt that some people have now and then, physics also teaches us, as we will see, that the matter of the "world" is much more complicated than it might seem at first. So we'd better be careful from the start and try to distinguish precisely (1) what is a compelling premise of natural science, (2) what its results are, and (3) what we *believe* independently of that.

The notion of a *common world* works because we learn that we can communicate about space and time and the phenomena we observe in them, and that the phenomena that others observe are essentially the same phenomena that I observe. This is a necessary condition for natural science. However, we also see that there are many areas in this world that are still fuzzy and unclear; there are "white spots" about which we know nothing yet, we suspect connections without being able to pin them down completely so far. Sometimes there are also subjective differences, cases in which we experience things in different ways or express ourselves differently about them. That is why we pursue natural science, in order to fill in the white spots, in order to achieve clarity, in order to recognize connections, in order to come to an agreement, in other words: in order to fully grasp and penetrate the world as a *common* world. It is this common world that is called *nature* in natural science (in contrast to the colloquial concept of nature, where nature is mostly meant in contrast to man-made; one thinks there of forests and meadows, lakes and rocks, animals in the wild, and so on), and the *knowledge* that natural science finds is about it.

But this world remains – it cannot be emphasized enough – a human imagination, a mental construct that is painted with human terms. The world is not the same as reality. This was made quite clear by Kant in the *Critique of Pure Reason* (we remember Chap. 2): space and matter *appear* only as something outside of us, but are in fact ideas *within* us.

Now, on the one hand, there are areas in the world that are directly accessible to our *senses,* that we can see, hear and touch. It is these areas that form our idea of *things*, things that have a color, size, shape, solidity and heaviness, that occupy a certain position in space, that have a certain history, all properties by which we can sensually imagine something. This includes the idea that things are *made* of something we call *matter*; and the idea we have of matter is shaped by the same sensory properties as our idea of things. In this way our idea of a world is concretized into an idea of a space with things in it, the things being constellations of matter. But then there are also *abstract* realms that only the natural sciences make accessible to us, especially physics – realms that are represented in abstract, largely mathematical terms, to which we have no sensory access, but which *constitute* hypotheses and theories that in turn efficiently describe connections between sensually accessible phenomena.

We now usually take it to mean that the abstract theories *explain* the sensually perceived phenomena, that is, that we can *understand the* latter in this way. Now what does this mean? Suppose we throw a stone obliquely into the air and watch it fall back to earth on a parabolic trajectory. We say Newton's law of gravitation *explains why* the stone falls to the ground and its trajectory has the aforementioned shape. By this we say that we accept the law of gravitation, an abstract equation, as a fundamental law of the world and that we understand *how* the trajectory follows logically from it. But in doing so, we do not understand *why* the law of gravity actually holds, we only see the logical connection between the law and the trajectory. We see how the specific phenomenon – the trajectory of the stone – follows from the general law. The general law, in turn, was deduced from the observation of many special trajectories.

But one can also see the matter in this way: The law of gravitation is the most efficient *description,* the most compact *summary of* the trajectories of freely falling bodies: if you want to summarize all the elliptical, parabolic, and hyperbolic trajectories of falling bodies and, at the same time, their velocities at any given moment in a single sentence, you have no choice but to express the law of gravitation. It compresses the entire information of these orbits to a minimum, to a single equation. And this is also how the aforementioned Ernst Mach sees the task of science: it is a matter of finding the most compact form of expression, the shortest possible description, which encompasses all observed phenomena. (Knowledge is thus to be *reduced in* a certain sense, not increased.) In this view, it is thus not a matter of explaining, but of describing. The difference is in what status of reality one ascribes to the abstract concepts that appear in the laws. In Mach's view, the concept of gravity is a mere tool, a concept that just helps us to summarize the trajectories of bodies in compact form, but not something to which one would ascribe an independent reality outside of this function. However, *if* we ascribe such a reality to gravity (or to the law it satisfies), only then does the law of gravity acquire an *explanatory* character and explain, for example, "why" the "really existing" gravitational field of the sun, which satisfies this "true" law, leads to elliptical planetary orbits.

Which view is now the correct one? I think that the second view is intuitively *appealing*. If the orbits of so many different bodies can be summed up in such a compact way, then it is only *natural to* ascribe a deeper meaning, a *reality*, to the concepts that accomplish this, which is then used as an *explanation*. (It is almost, but not quite, as natural as ascribing a reality to space, matter, and things, which are, after all, initially also only conceptions, albeit ones that are much more directly accessible to us). This, however, is a *feeling*, not something that can be proved. However, the feeling is put to the test by the fact that the concepts to which one would like to ascribe reality often turn out in retrospect to have only a limited range of validity. Thus, in the case of the law of gravitation, it turns out that it becomes inaccurate in certain situations, and even that the concept of gravity itself is no longer applicable there, but must be replaced by another concept, that of the curvature of spacetime. We will come back to this.

At the end of the nineteenth century, Mach insisted on his view, especially in the case of atoms. At that time, it had become apparent that both thermodynamics (heat theory) and chemistry could best be *explained* by the fact that all matter consists of atoms or molecules. Mach insisted on not ascribing any real existence to atoms, but only using them as an abstract concept, as a mental tool with which precisely the most compact, efficient *description of* the observed phenomena could be achieved. But when, at the beginning of the twentieth century, atoms could be observed and measured in a much more direct way, their status as reality was thereby raised to such an extent that Mach's view was thus discredited. However, looking at the internal structure of atoms, one ends up with quantum mechanics, which has properties so strange that Mach's view becomes more appealing again. We'll come back to that in detail as well.

Whether we understand natural laws in terms of an *explanation* or a *description* has no consequence for the actual operation of natural science, and certainly not for the technical applications we construct from its results. The difference plays at most a role in our personal motivation why we do natural science and how we interpret its results.

Another difficult issue arises when we apply natural science to ourselves, in particular to our so-called *cognitive apparatus,* that part of us which is responsible for gaining knowledge, namely the senses and the mind. What can we know about our cognition in a scientific way? A scientific approach presupposes that we regard the cognitive apparatus as something within the world, that is, as an objective phenomenon given in the form of material things in space. In this case, these things are the sense organs and the brain. So we study the behavior of the sense organs and the brain and thus obtain scientific knowledge about the process of cognition.

The prejudice that the world is the same as reality leads many to the assumption that this knowledge can completely encompass the cognitive apparatus. In the sense of Schrödinger's dilemma (Chap. 2), however, one can also see it in such a way that the cognitive apparatus comes *before* the world from a purely logical point of view. It is precisely the cognitive apparatus that has constructed this world as a conception, and the world is initially nothing other than this conception, this construction. The success of natural science shows how much of this world can actually be objectified. Through this objectification the world seems to be lifted above its status as an imagination, it appears as a reality existing independently of us, and this independent reality is also accepted by a large majority, especially among natural scientists. But if one looks closely (and we will do so in the course of this book), this assumption leads to some inconsistencies and needs to be reconsidered carefully.

The way of natural science has proved to be tremendously successful and powerful. It *works in* a great way. So we clearly did something *right* when we chose this path. The picture that natural science paints for us is magnificent, complex, and beautiful (even if some backward-looking nostalgists would have us believe otherwise). As philosophers, however, we need to be somewhat careful about how we interpret this success and what status of reality we ascribe to this picture. Natural science takes place on the level of objectivity, not reality, and therefore has limits.

5

Reductionism

What is remarkable about the natural sciences is that each contribution is placed as a building block in the whole. All the millions of technical articles, experiments, observational data, textbooks, etc., together form what I will call the **picture of the natural sciences,** a cathedral of knowledge on a vast scale. So vast that by now almost all phenomena and objects we encounter in the world can be classified with their help and explained in a certain sense (see above).

Natural science is divided into several, partly overlapping fields (e.g. biology, chemistry, physics), which in turn are divided into subfields (e.g. zoology, botany, molecular biology). Within physics we can subdivide either by subfields or by theories (this is a special feature of physics). Subfields depend on the class of phenomena or objects to be studied: Astrophysics deals with celestial bodies, solid-state physics with solids of all kinds, hydrodynamics with liquids and gases, atomic physics with atoms, and particle physics with elementary particles (for atoms, contrary to their name, which means "indivisible" in Greek, are after all divisible into even smaller particles). Nuclear physics deals with atomic nuclei (smaller than an atom, but still composed of several particles), thermodynamics with thermal phenomena, gravitational physics with gravity, electromagnetism with electrical and magnetic phenomena including light, and so on. These subfields are not always strictly demarcated from each other. Especially in astrophysics almost all other fields play a role: For example, one needs hydrodynamics, nuclear physics and thermodynamics to understand the interior of stars, gravitational physics for the movements of celestial bodies relative to each other, and so on.

© Springer-Verlag GmbH Germany, part of Springer Nature 2022
J.-M. Schwindt, *Universe Without Things*,
https://doi.org/10.1007/978-3-662-65426-2_5

On the other hand, there are the theories, mathematical formalisms that can be used to quantitatively describe and predict large classes of phenomena. Quantum field theory (QFT), for example, is the mathematical formalism used to treat particle physics. Maxwell's theory describes electromagnetism. For gravitation there are two theories: Newton's theory and Einstein's theory of gravitation. The latter is known as the General Theory of Relativity (GR). It is more accurate than Newton's, applies in a greater number of cases, and describes a greater number of phenomena (including, for example, gravitational waves). Newton's theory, on the other hand, is much simpler and quite sufficient for most gravitational phenomena we deal with in everyday life.

What physical theories actually are and do – both in general and the theories just mentioned in particular – will be discussed later in the book. In this chapter, which is about reductionism, I will anticipate some properties of these theories in examples. This will probably leave some questions unanswered for some readers. Please just read the rest of the book to clarify them. It seemed appropriate to me to place the reductionism chapter before the explanation of the theories, because it deals with the structure of the natural sciences as a whole and with the special role that physics plays in it. The reader should already have this special feature in mind when we deal with the theories in detail later.

The number of physical theories is quite small and manageable. If you want to study a particular physical phenomenon theoretically, you usually only have to answer a few yes/no questions to find out which theory to apply. For example: Is the situation better described by individual objects or by a continuum? If the former, is the number of objects very large? Do velocities near the speed of light occur? Is quantum behavior expected to play a role?

Within their respective spheres of competence, some theories have been backed up by so many experiments and with so much precision that they are regarded as instances of the highest authority. They are not, therefore, theories which still have the character of mere hypotheses or speculations, but have withstood constant critical scrutiny to such an extent that the physicist consults them as the judge consults his law-book. These theories I call the **confirmed** theories. Their respective spheres of competence are known, so far as they lie within the limits accessible to us by experiments or astronomical observations. Beyond that – for example, at very high energies that we have not yet been able to generate with our particle accelerators – certainty ends, and theories that refer to these ranges have the character of hypotheses. This is all quite roughly speaking, and abbreviated as much as is necessary for the context of this chapter. Later on we will specify all this further.

Thanks to the small number of theories, physics can be divided into theories instead of phenomenological subfields. In the study of physics, one takes courses in theoretical and experimental physics. Naturally, the lectures of experimental physics are rather subdivided according to phenomenological subareas, the lectures of theoretical physics rather according to theories.

Let us compare this with the situation in biology. There are no comparable theories there. Imagine, for example, a theoretical zoology in which the behavior and structure of all animal species could be predicted from a few assumptions with a precise formalism. The diversity in biology is simply far too great. In physics, we have to deal with only four basic forces that determine the physical behaviour of all things, and with a few types of particles whose properties can be summarised in a few mathematical parameters. In biology, on the other hand, we are faced with millions of animal and plant species, billions of different biomolecules. The task here is rather to catalogue and classify, to describe behaviour and to find individual causal chains. Sometimes it is possible to identify more general patterns and relationships. This already leads to the establishment of generally valid theories, such as the theory of evolution or genetics. In practice, however, these tend to operate in the background as general principles and explanatory patterns, without offering the possibility of predicting the development of living beings over a longer period of time. The general theory of genetics, for example, does not allow us to predict which genes of a particular organism are responsible for which external characteristics. This requires a laborious cataloguing of individual observations. In biology, the closest thing to physical theories in terms of precision is the description of individual mechanisms where their functioning is understood chemically and physically, e.g. the electrochemical transmission of impulses between neurons (nerve cells) in the brain, the chemical processes involved in certain metabolic processes and the reproduction of DNA. However, since each of these theories is only responsible for a very small area of the whole of biology, their number is very large and also does not have the same "area coverage" as the theories in physics.

This difference also determines the study of the two subjects. In biology as well as in medicine, you have to learn an incredible amount of information by heart, the structure of organs, the course of individual veins and nerve tracts, countless chemical substances, their effects and reactions, causal chains. In physics, on the other hand, knowledge of a few equations and principles is sufficient. The difficulty in studying physics is to acquire the mathematical "tools" necessary to understand these equations and draw the correct conclusions from them. In a typical oral exam, you will be asked to first write down the basic equations and then derive and explain certain relationships from them. In a typical written exam, you will be given the task of determining the numerical values for some physical quantities from a given physical situation with given numerical values for other physical quantities. Again, you will first write down the basic equations, transform them appropriately for the purpose of the task, and then insert the numerical values. The art here is not in factual knowledge, but in reasoning.

What about chemistry? Chemistry is the study of chemical substances, their properties and methods of producing them through chemical reactions. A chemical substance is characterized by its smallest units: Atoms, molecules, ions. Water, for example, is the substance composed of water molecules; these are molecules composed of two hydrogen atoms and one oxygen atom, in chemical shorthand H_2O. Substances whose smallest units are atoms are called elements. In the middle of the nineteenth century it was discovered that the elements can be numbered and divided into groups according to a certain scheme, schematically represented in the famous periodic table of the elements. At that time, nothing was known about the structure of the atoms, so the classification was based solely on their behaviour in chemical reactions. At the beginning of the twentieth century, however, it became clear that theory in chemistry is pure physics.

All chemistry can be theoretically traced back to quantum mechanics (QM) and thermodynamics. QM, applied to positively charged atomic nuclei and negatively charged electrons, determines (and in a sense "explains") the structure of atoms. The number of an element in the periodic table corresponds to the number of electrons in the atom. The division of elements into periods and groups follows from the way the electrons in the atom group themselves spatially according to the laws of quantum mechanics. Several atoms (the same or different) can share a certain number of their electrons, which makes these atoms "bound" to each other. These are the molecules of chemical compounds. Or an electron passes entirely from one atom to another. This gives the atoms an electric charge and they become ions. All this follows from the laws of QM. QM also determines the energy values of each state and thus how much energy is released or must be added in a chemical reaction. Furthermore, it describes how the atoms in a molecule are geometrically arranged and how the electrical charges are exactly distributed in it. Thus, it ultimately determines all the properties of the chemical substance. In the case of water, for example, it tells us that the electrons, i.e. the negative charge, are located somewhat closer to the oxygen atom than to the hydrogen atoms. This creates electrical forces between the molecules, called hydrogen bonds. From these forces and the property (which also follows from QM) that the three atoms of the water molecule are arranged at an angle of 104°, all the properties of water ultimately result: melting and boiling behaviour, shape of the ice crystals, thermal conductivity, surface tension, etc.

Thermodynamics, on the other hand, explains those aspects of chemical reactions that are not already determined by QM. The energy difference between two bond states (unbound atoms vs. atoms bound in the molecule) follows as said from QM. In chemical reactions, however, the kinetic energy of the atoms or molecules must also be taken into account for the energy balance. This in turn is related to the temperature. In addition, the pressure under which a reaction takes place plays a role. Technically, such external conditions (pressure, temperature) can be controlled to a certain extent. The effects of these conditions on the reaction behaviour are described by thermodynamics.

It can therefore be said that chemistry in theory was traced back to physics. In the mid-nineteenth century, chemists could still justifiably claim to have unlocked nature's innermost secrets with their research. The establishment of the periodic table was a tremendous scientific achievement. However, after the connection with QM became clear 60 years later, chemistry lost its status as an actual fundamental science. The chemistry of today is a purpose-driven science. (It was that to a large extent 150 years ago too, but then it was still basic science). Its main task is to develop efficient processes to produce or isolate substances that are useful to humans or the environment and to weed out harmful ones. It also helps biology understand its fundamentals. If you look at what the Nobel Prizes in Chemistry have been awarded for over the last few decades, you will find that they are almost exclusively split between these two areas: Development of technical processes and findings in the field of biochemistry.

The reduction of chemistry to physics does not mean that everything is clear. The basic equations of physics all have a pitfall: They are relatively easy to write down and understand (once you have acquired the mathematical basics), but difficult to solve when more than two objects are involved. This is already true for gravity: Newton's law of gravitation is extremely simple, a simple equation that sets the force as a function of masses and distances. To "solve" the equation is to calculate, for a given situation, the trajectories of objects that behave according to that equation. Solving the equation is quite simple when only two objects are involved: a rock falling to Earth; a planet moving around the Sun. The solutions of such special cases can themselves be put back into simple equations. Because these solutions (here: the trajectories of stones and planets) can be directly observed in nature, they are often discovered before the actual law (here: Newton's law of gravity) of which they are the solution. Galileo recorded the trajectory of a falling stone in the early seventeenth century, now known as "Galileo's law of falling bodies". At about the same time, Kepler established his three laws describing planetary orbits. Then, at the end of the seventeenth century, Newton found the law of gravitation, which is, in a sense, the "cause" of these rules.

But as soon as more than two objects are involved (it starts with the so-called three-body problem), the whole thing becomes very complicated at once. The basic equation remains as simple as it was, but the solutions no longer have a simple form and can only be approximated. The problem is not so great when the effect of one object very much outweighs the others. In principle, after all, the solar system is a nine-body problem. Eight planets move around a sun. But the Sun exerts a much stronger force on the planets than the planets exert on each other that the latter cause only small deviations from Keplerian orbits (though these will be noticeable in the long run). But if the forces are of roughly the same magnitude, you'll quickly end up in hot water.

We have such a situation in chemistry: The electric forces between electrons are of the same order of magnitude as the forces between electrons and atomic nuclei and those between atomic nuclei among themselves. It is utterly hopeless to try to solve the relatively simple fundamental equation (in this case, the stationary Schrödinger equation, a fundamental equation of QM) exactly. In this case, "solving" means to determine from the equation the geometric arrangement of the molecules and therein the spatial distribution of the electron shells. This task can only be tackled with tricky approximation methods and requires intensive use of computers. The development of such methods is the task of theoretical chemistry (also known as quantum chemistry). I would like to emphasize that this is also an applied problem, not a fundamental problem. The theoretical foundations are well understood. The problem is to design computational methods that can be used in practice to efficiently infer the needed conclusions (the "solutions") from the known fundamentals. For relatively small molecules this has been achieved, but for large ones, especially the complex molecules of biochemistry, progress has been rather slow. Therefore, chemistry in these areas continues to draw its conclusions from experiment and observation, not theory.

I would also like to emphasize (and this is important for the further discussion of reductionism) that this restriction is by no means a reason to doubt the validity of QM for large molecules. After all, you don't doubt the validity of gravity merely because you are unable to solve a general nine-body problem for practical-computational reasons. QM, just like Newtonian gravity, is one of the confirmed theories that has been proven in millions of examples and critical tests. We must accept that calculations are often difficult, and should be pleased that at least the theoretical foundations on which these calculations are based are understood.

The reduction (backtracking) of chemistry to physics is an example of a **reduction in principle**. That is, we have understood in principle and confirmed by many examples how it works, but for practical-computational reasons we cannot perform the derivations (here: the derivation of molecular structures from QM) for the more complicated cases.

In general, in the natural sciences, some areas, sub-areas and theories are more "fundamental" than others. There is a hierarchy. Where is up and down in a hierarchy is a matter of convention. In hierarchies of people, the boss is "up" and the subordinates are "down". In hierarchies of scientific fields, subfields, and theories, we imagine the hierarchy to be reversed: The more general, more powerful theory is "at the bottom." This is perhaps because the natural scientist, like a deep-sea diver, dives deeper and deeper into objective reality, moves further and further away from the surface of phenomena, seems to get closer and closer to the bottom of reality the further he sinks into the general and abstract, where our everyday mind can hardly see any light.

The deepest (i.e. most general, most powerful) known theory in this hierarchy of natural sciences is QFT. Whether this in turn can be traced back to an even more fundamental theory is not known. In any case, so far there are no observations that force us to introduce such a theory. (Some would probably place GR on the same level as QFT, and note that the two must be merged at an even deeper level. However, this is controversial. My view is that there is some evidence to suggest that GR can also be traced back to a variant of QFT, i.e. is positioned one level above the latter).

Now, the hierarchy of natural sciences is such that from every starting point, in a sense yet to be clarified, ways of reduction downwards can be found until one has arrived at the level of QFT (only in the case of GR this is, as said, controversial). Conversely, in a sense still to be clarified, one can also say that from QFT all higher levels follow logically, **in principle**, just as chemistry follows in principle from QM and thermodynamics.

The hypothesis that this is so is called **reductionism**. But reductionism can be understood in different ways, and that is why I have twice added the phrase "in a sense yet to be clarified".

There is a fairly popular view that can be summed up with the slogan *"The whole is greater than the sum of its parts."* In the terminology of philosophical isms, this is called holism or emergentism. This current of thought sees itself as the opposite of reductionism, i.e. it assumes that the sentence *"The whole is more than the sum of its parts"* *is* in contradiction to reductionism. But does reductionism even say that the whole is equal to the sum of its parts? For that, we now need to discuss the "meaning yet to be clarified". So let's look at what different forms reduction can take.

5.1   Reduction by Decomposition

The best known form of scientific reduction is that which leads down to elementary particles by decomposition of macroscopic objects in several steps. It is most clearly represented in the biology-chemistry-physics chain. We can understand a human or animal body from the interaction of its organs. The possible movements, for example, result from the interconnections of muscles, bones and joints, the overall process of digestion from the interconnections of the individual digestive organs. The organs in turn consist of specialized cells whose interaction determines the functioning of the organ. The cells consist of organelles (small "organs" into which the cells can be subdivided), and these in turn consist of a large number of different substances, including huge biomolecules with tens of thousands (in the case of proteins) to several billion (in the case of chromosomes) atoms. The biomolecules can often be subdivided again into smaller units in several steps. Human chromosomes, for example, each contain up to 1500 genes and up to 250 million base pairs. The four bases adenine, cytosine, guanine and thymine each consist of about 15 atoms. The atoms consist of negatively charged electrons in the "shell" and the much smaller nucleus. The nucleus is made up of protons and neutrons. Protons and neutrons "consist of" quarks, but in this last step we are in a borderland where the term "consists of" is no longer entirely justified.

Each level of this **decomposition hierarchy** has its own terminology and methodology. The movements of a human or animal as well as the "input/output behaviour" of its digestion, i.e. its food intake and excretions, can still be observed directly. The internal organs are first known from the dissection of cadavers. For the examination of cells one needs a microscope. The structure of molecules can still be seen with electron microscopes, which are microscopes that work with electron beams instead of light. In the subatomic range, scattering experiments are the preferred method, and structures are revealed indirectly, with the help of mathematical models.

Splitting objects into smaller objects also works very differently depending on where you are in the hierarchy. Objects on the upper level can still be cut up with a scalpel. Molecules can be attacked with heat or chemical processes. In order to split a larger atomic nucleus, a so-called nuclear fission, it must in many cases be bombarded with smaller particles.

As a rule, however, you cannot understand an object by cutting it up. The decisive thing about this form of reductionism is precisely the **interaction of the individual parts**. Placing the individual parts neatly next to each other is of no use at all, because what matters is how they are linked to each other, how they interact with each other and how they play together in the overall picture. In this sense, it is obvious that the whole is more than the sum of its parts, and this is the only way reductionism can work. The statement for this reduction is then: **one can trace the behaviour of an object back to the behaviour of its individual parts and their interactions with each other**.

By "tracing back" is meant: The behavior of the object is thereby causally and logically explained, i.e., the behavior of the individual parts and their interactions determine the behavior of the total object. Every structure, every rule, every quantity, which belongs to the total object, is logically and causally based on the structures, rules, quantities of the individual parts and on the structures, rules, quantities of the interactions. A why-question on the level of the total object has a because-answer on the level of the individual parts.

The interactions are quite crucial for these explanatory patterns. Interactions can make the qualitative and quantitative behaviour of the whole object look quite different from that of the individual parts. In a sense, "something new" is added. This is called **weak emergence**. In contrast, some philosophers and natural scientists argue for **strong emergence**, which means that the new thing that emerges at the higher level cannot in principle be traced back to the lower level, i.e. that of the individual parts and their interactions.

One can express the difference between the two views by the level at which the interactions are to be located. In the case of weak emergence, we assume that the interactions are to be understood at the level of the individual parts. That is, we know, for example, what forces are at play when two individual parts come close to each other, whether they attract or repel, and how these forces behave as a function of distance and spatial orientation. We also know how to sum up the forces in the case of more than two parts. Or in the case of biological objects, we know what the surfaces of the individual parts (e.g., cells) are like, what substances can and cannot pass through these surfaces, to what extent the surfaces are elastic and deformable, etc., and from this we know how two adjacent individual parts interact with each other. In the case of strong emergence, it is assumed that the parts in the overall object interact in a way that cannot, in principle, be understood as a summation of such individual interactions, but represent something entirely new – an interaction that can only be grasped and described at the level of the overall object.

A relatively well-known example of a theory of strong emergence was nineteenth century vitalism. At that time, no one could imagine how life would arise from an interplay of chemical processes. Therefore, vitalism assumed that in the case of living beings an additional force is added to the natural forces, the *elan vital*, a force that occurs only in such constellations of matter as are given in the form of living beings, and cannot in any way be attributed to an interplay of more elementary forces.

Such a strong emergence seems illogical to me, not only in the case of vitalism (which is considered long outdated), but in general. Lastly, all materia is composed of elementary particles. Their interactions are precisely determined mathematically, no matter how many of them there are. We know that in certain cases these interactions lead to chain reactions in which very many particles interact: to the formation of macroscopic structures, of stars, planets, clouds and crystals. The interaction of biomolecules has also been explored to a large extent. We also know, as described in the example of chemistry, that there are practical limits to computability. Thus, if we know of a macroscopic phenomenon that could not previously be deduced from microscopic rules, we should first attribute this to these practical limits of computability, and not to the failure of reductionism.

Moreover, the addition of something new in the sense of strong emergence means a contradiction to the existing laws of nature. Since the interactions of the particles are already fixed by the latter, the new thing must somehow modify these interactions, i.e. contradict the known rules. Now it is not impossible that microscopic rules may have to be changed in certain cases. New particles are discovered that interact with already known particles; mathematical rules for interactions are refined. But in these cases it is a modification of the microscopic laws of nature, that is, at the level of particles, in the course of the progress of science. Strong emergence, on the other hand, wants macroscopic behavior to change, to exhibit something new that does not follow from the microscopic rules. But neither should it override or modify those microscopic rules. This is a contradiction precisely because the microscopic rules already determine everything.

Weak emergence, by the way, does not only occur in macroscopic systems, but already in the range of the smallest particles. For example, we can consider the mass of the proton as emergent. Above I had written that a proton consists of three quarks, but that the designation "consists of" is no longer completely justified here. The proton is 100 times heavier than the three quarks together. This is because the quarks interact so strongly with each other that a whole cocktail of so-called virtual particles is created, which scurry around in the proton and contribute to its mass. Because of the complexity of the effects that occur, this mass is very difficult to calculate from the properties of the quarks and their interactions. It virtually "emerges" from the scurry of virtual particles. Experimentally, the mass of the proton has been known since its discovery 100 years ago. That it consists of quarks has been known since the 1970s. But it was not until 2008 that a group of researchers spread across half of Europe succeeded, with the help of supercomputers, in calculating the mass of the proton from the interactions of the quarks, fortunately in very good agreement with the value known from experiments.

At the next level up, protons and their electrically neutral siblings, the neutrons, combine to form atomic nuclei, which again have properties that are difficult to calculate from the parts, although of course they follow mathematically from them in principle. Next, the atomic nuclei combine with electrons to form atoms. This step is comparatively easy because only electric and, to a small extent, magnetic forces are involved, which are much simpler than the nuclear forces in the atomic nucleus and proton. And because the atomic nucleus is so much heavier than the electrons that it can be considered, to a good approximation, as an immobile "central star" at the center of the atom, like the sun in the planetary system. We are one level further on in chemistry, where atoms combine to form molecules or ionic compounds. Here, a symmetry is lost that was still present at the lower levels: protons, atomic nuclei, and atoms are all spherical, at least as long as there are no external forces acting on them. Molecules, on the other hand, form intricate geometric patterns, and this is what accounts for their difficulty. These patterns lead to new "emergent" properties and behaviors that are not found in the parts, i.e. atoms, and are not so easily calculated from them. Think, for example, of the complicated hexagonal patterns of snowflakes, which ultimately arise from the geometry of the water molecule. Or the folding of proteins, which determines their entire mode of action in biochemistry. Or the double helix structure of DNA. Finally, at the level of biology, we find all the "emergent" properties and behaviors that we call life: Reproduction, heredity, metabolism, response to stimuli from the environment, etc.

It is this form of reductionism that holds the sciences together; that ensures that with physics we can secure chemistry and with chemistry we can secure biology from below, give it a foundation. In dealing with it, however, we also encounter many difficulties which relativize its practical value for us, but also make its character clear:

- In many cases, reduction is not accompanied by real **understanding.** It is just a reduction in *principle,* i.e., we know the mechanisms that are involved, but *how* the phenomenon under investigation follows from them, we often do not know, only *that* it is so. In some cases, such as the proton, we can do the calculation with supercomputers, but without getting the "aha" moment we wish for. In other cases, even the best supercomputers are not enough. I think this is quite normal and does not argue against reductionism. We can't expect every phenomenon to delight us, after in-depth analysis, with an aha moment like a whodunit. We can only strive to understand *as much as practically possible.*

- Often the reduction is not from one level to the next, but to several lower levels at once. To explain the functioning of the eye, we need to know the properties of light, understand the optics of the lens and the biochemistry and biophysics of the sensory cells. So we refer to regularities that come from levels of biology, chemistry and physics at the same time. Radioactive decays in nature, phenomena from nuclear physics, play into genetics as they contribute to the origin of mutations. The weather phenomena studied in meteorology involve a variety of chemical and physical, and in some cases even biological, processes. In the case of such an interplay of several levels, the reduction can be continued in a cascade-like manner until the system under investigation is completely reduced to a common level, in extreme cases that of QFT or particle physics.

- Every system is subject to external and internal forces. To understand a system in terms of reductionism from its parts and their interactions, on the other hand, requires that we be able to view it as an isolated system whose behavior follows solely from its internal forces, or at least to control the *external influences* so that they do not form a significant complication of the event under investigation. In physics, with the aid of many techniques, it is often possible enough to separate the internal and external influences and isolate the desired effect so that reduction "from within" is possible. In chemistry, under laboratory conditions, the external conditions (e.g. constant temperature, constant pressure) can be designed in such a way that a reaction under investigation is not without external influences, but these are only included in the calculations as simple constants. In biology, too, it is possible to control external conditions in the laboratory, for example, when a cell culture is being studied there. But if, for example, the behaviour of living beings "in the wild" is studied, then the number of external influences is uncontrollably large. The behaviour of living beings can then no longer be explained individually, but only in combination with internal behaviour and the interactions with the rest of the ecosystem. This is perhaps the biggest problem of reductionism: each whole is again part of an even larger whole with which it interacts. So reductionism is ultimately only completely accurate if you include the *whole universe*. Now this is a very big form of "in principle".

In the practice of science, of course, this is completely utopian. We should therefore understand reductionism in application rather practically: In many cases we can estimate approximately how large the influence of the environment on a given process is. If it is small, we can carry out the reduction within the system concerned. Nature itself makes it somewhat easier for us in many cases, because it produces stable systems of its own accord that make themselves as free as possible from external influences by surrounding themselves with protective layers and using compensatory mechanisms. For example, our body keeps its temperature at a constant 37 °C, whether the outside temperature is 0 °C or 40°. In many cases, the external influences are of such different orders of magnitude that only a few are relevant in practice. Of the external influences on planet Earth, for example, the most important are the gravitational forces of the Sun and Moon and the insolation of sunlight, which are relatively easy to quantify. You can disregard other effects, such as light from the stars or cosmic rays, in most cases when studying processes on Earth. Unless, (1) you are studying a special effect in which one of these smaller influences plays a role, or, (2) a major meteorite with global consequences strikes the Earth, as happens every few million years; this is a large but just rare effect.

• Reductionism explains to us in *principle*, by means of the genetic code, why certain genes lead to certain characteristics of an individual. But it does not explain to us why exactly this combination of genes is realized at all, in particular why man has developed just in this way and not differently. The theory of evolution in combination with genetics explains to us the mechanisms by which development takes place. But in the concrete development that took place on earth, **coincidences** play a decisive role – things that could have happened one way or another (philosophers speak here of **contingency**). How does reductionism deal with these coincidences?

In fact, we must distinguish between **the reduction of patterns** and **the reduction of developments.** Here, by patterns, I mean everything that can be said in general terms without reference to temporal processes in concretely realized individual systems. For example: Water has such and such density, surface tension, electrical conductivity etc. at such and such temperature. Or: This gene leads to the synthesis of this protein with these properties, which has these consequences in an organism under these conditions. Or: The effect of a moon's gravity on its associated planet leads to tides on that planet that can be calculated this way and that. The sections so far have been mainly concerned with the reduction of patterns. In the case of developments, the matter becomes even more complicated: Here, not only the general laws of nature must be taken into account, but also the exact initial conditions of the concrete system to which the rules are then to be applied. Tiny differences at the beginning can lead to huge differences in the long run. This problem has become known as the **butterfly effect**: The flap of a butterfly's wings in Brazil can determine whether a tornado will form in Texas a few days later. (This is why our weather forecasts are so bad when they try to look more than a week into the future). So we would need to know the exact positions and states of all the molecules on Earth at an early date to be able to in *principle* deduce the evolution from that. But even that would not be enough. Because of the butterfly effect, the smaller external influences on the earth now also play a role, e.g. which cosmic particle hits the earth and when exactly. In addition, according to quantum physics, there is a randomness that in principle cannot be resolved further: When exactly a certain radioactive atomic nucleus decays, for example, is subject to absolute chance. In principle, physics can only provide half-lives, i.e. when statistically about half of the atomic nuclei of a species have decayed, but can say nothing about the individual fates of the individual nuclei. The decay of a particular individual nucleus can, however, be decisive in determining whether or not a particular genetic mutation is triggered, with potentially far-reaching consequences. Thus, the specific long-term evolution of systems cannot be fully explained from reductionism. This is another limitation, in addition to those from the previous points.

Nevertheless, reductionism can also be applied to developments if we understand it as a **reductionism of causal connections**. When we explore the history of the earth, we come across many things that seem random to us, but also many causal connections. Causal connections are subject to general laws of nature (patterns), so the reductionism of patterns comes into play again. For example, we assume that the impact of a certain meteorite 65 million years ago caused the extinction of the dinosaurs. This causal relationship can now be analyzed and thus made plausible. The impact itself is subject to the laws of physics, as are the direct consequences: the formation of the crater, the shock wave, the matter thrown into the atmosphere, and the heat generated. Numerous meteorological effects are the result, extreme weather conditions, which in turn have a biological impact, so much so that they lead to the extinction of many species.

- The hierarchy of sciences can be continued above biology, e.g. to sociology. Sociology is now already not really counted among the natural sciences, although its methods are of course scientific. The problem is that the regularities it finds are not very general. They mostly relate to a concrete society at a concrete point in time and are not easily transferable. Still, we can use them to better understand the concrete state of our society, to see connections in it, and to infer possible actions that can be taken to improve certain conditions. A reduction of sociology would mean reducing the overall interrelationships of society to the interaction of individuals; in essence, then, this would be a reduction to psychology. Regularities of society would then be understood as the logical consequence of the psychological regularities of human beings. However, the external conditions to be considered are much more extensive than in the "lower" sciences. In chemical reactions, ideally, you only need to know the externally imposed temperature and pressure; everything else follows from the internal dynamics of the reaction. In the case of social processes, on the other hand, you have to take into account the entire highly complex cultural context, economic, legal and political framework conditions and, in some cases, even climate changes. This gigantic "background ballast" severely limits the applicability of reduction, especially if it is supposed to be quantitative, i.e. to make statements about *how many* people follow a certain trend, respond to a certain measure, etc. Only in the case of some simple patterns, where certain economic incentives are given to certain people within society to do a certain thing, can the relationship between psychological basis and sociological consequence be made reasonably precise. Qualitative correlations, such as why populations grow more in some countries than others, are much more common.

The reductionist statement that, after all, the development of society follows quantitatively exactly from the behavior of the atoms and molecules of which it is ultimately composed, while true, is utterly useless to the sociologist. But sociology, as a science, is highly relevant to political decisions, perhaps more so than any other science. This shows that reductionism is largely irrelevant to such issues.

- We are particularly sensitive to the issue of reductionism when it comes to ourselves as individuals. It is one of the longest and most heated discussions in the border area of philosophy and natural science to what extent human consciousness can be traced back to brain processes. The subjective character of our experience is so very different from the objectively measurable electrochemical behavior of brain cells. My consciousness feels like an indivisible entity that cannot be composed of interactions of pieces of it. Our mental inner life seems to be much more than just a fully automatic firework of nerve impulses. Moreover, we have the strong feeling of having free will, i.e. of determining to a large extent our actions ourselves, without these decisions being predetermined from the strict mathematical laws to which matter is subject.

Does reductionism not apply here? This is a very difficult issue, and we will return to it at length in Chap. 11. My position on this is something like this: The natural sciences are reductionist by their very nature, in all cases. Psychology is no exception. With great success, many of our cognitive abilities have already been traced back to the workings of the brain. Even more, we have succeeded in producing certain forms of artificial intelligence in computers using algorithms that resemble the way our brain cells (neurons) work, and are therefore called neural networks. We have taught computers to compose music. We even know which region of the brain to manipulate to induce something like spiritual experiences in a human being. On free will, there are experiments that show us that we are often deluded about when and why we made a decision (we'll come back to this). These findings are certainly challenging to our self-esteem. This seems to be the price to pay for a complete scientific penetration of nature.

However, there is a big but: it is the natural sciences that have this reductionist character, not necessarily reality itself. The natural sciences represent a certain point of view on reality, defined by their methods, as described in Chap. 4. In particular, this point of view is oriented towards objectivity. I think it is quite possible that our subjective experience includes aspects of reality that are "invisible" to this method and whose character need not be reductionist. Perhaps within the framework of these aspects there is even a concept of freedom that takes place outside of such simple decision-making schemes as are known from psychological experiments. But this is speculation, and I do not wish to dwell on it here. My point is: *if* any aspect of our subjective experience can be expressed in scientific terms, then reductionism is also applicable to it.

Let's summarize: Reduction by decomposition is a universal principle of the natural sciences. It states that the behaviour of a system follows logically from the behaviour of its individual parts and their interactions, as well as from the external influences to which the system is exposed. This leads to a logical hierarchy of scientific descriptions, at the bottom of which is the physics of elementary particles. The reduction also involves a reduction of causal relations, i.e., the causal relations at the upper levels of the hierarchy follow logically from the causal relations at the lower levels. It is a reduction *in principle*, i.e., it is clearly defined *how* the reduction is to be carried out in principle, but this *how* often fails because of the practical limits of computability. In other words: In the overall building of scientific theories, the principle of reduction by decomposition always expresses a truth. In the practice of scientific research, however, this reduction is sometimes applicable with success and sometimes not. Moreover, in practice, reduction is usually only possible in the form of approximations, i.e. certain effects must be neglected. Depending on the problem and complexity, this can lead to very exact quantitative or only rough qualitative statements. An absolutely exact reduction is only possible if the system and all external influences on it are completely reduced to the level of elementary particles.

5.2 Reduction by Generalization or Unification

In this form of reduction, several laws of nature are reduced to a single, more general law of nature. The individual laws then follow logically from the general law. By finding such general laws, the unification of physics in particular is advanced. An example has already been given: Galileo's law of falling bodies and Kepler's laws of planetary orbits both follow from Newton's law of gravitation. In the nineteenth century, Maxwell reduced the laws of electricity, magnetism, and light propagation all to his electromagnetic theory. In turn, in the 1960s, Maxwell's theory was unified with the theory of the weak nuclear force (responsible for radioactive decays) to form the theory of the "electroweak" interaction. GR generalizes the Special Theory of Relativity (SR) and also describes gravity.

It is a great stroke of luck for us that this form of reduction is so often possible. One could also imagine a world which even at the level of elementary particles (i.e. after the reduction by decomposition has already been carried out) still has a great many incoherent individual laws. But this is not the case. Through the unification it becomes obvious that there are only very few fundamental laws of nature, from which the others logically follow.

In contrast to reduction by decomposition, reduction by unification is *always* associated with a very great "aha" experience and a deeper understanding. It represents to the highest degree what we perceive as "beautiful" in natural science. Another difference between reduction by decomposition and reduction by unification is that the former always leads from the large to the small, from the macroscopic to the microscopic, but the latter does not. Instead, it leads from the particular to the general, without referring to proportions and part/whole relations.

5.3   Reduction by Replacement

Newton's theory of gravity can be reduced to Einstein's (i.e. to GR). But in doing so, the character of gravity is completely changed; one can say that Newtonian gravity is replaced by something completely different. With Newton, gravity is a force field, similar to the electric one, except that it is always attractive, whereas the electric force can be attractive or repulsive. With Einstein, on the other hand, gravity is not a force field, but an effect of the geometry of spacetime. This geometry is "curved", and this curvature causes an object to move as *if* it *were* made to do so by a gravitational force.

In this case, I speak of reduction by replacement. A "thing" or a phenomenon turns out to be something quite different from what it seemed to be. Something similar is also at the beginning of natural science: an event that previously looked like an act of the wrath of the gods (a thunderstorm, for example) suddenly finds an explanation within the framework of the laws of nature (lightning is a phenomenon of electrical discharges). But there is a difference: the wrath-of-the-gods theory was subsequently dropped (well, by most) because it no longer has any value to the rational. It has been shown to be false. Newton's theory, on the other hand, continues to be correct.

This is a point that is important to me. You sometimes find in popular science books the statement that Newton's theory is wrong, that Einstein has shown that Newton was wrong. You can't let that stand like that. Einstein's theory is more general and more precise than Newton's, but in most situations Newton's theory is quite sufficient and gives sufficiently precise results. With physical theories, we must generally assume that their range of validity is limited, that we must regard them as approximations that only yield accurate results under certain conditions. Where "accurate" means that, within the limits of our measurement accuracy, we cannot detect any deviation between the behavior predicted by theory and that found experimentally. Einstein's theory also has a limited range of validity; in order to describe gravity under extreme conditions (such as those found in black holes or near the Big Bang), it must be replaced by a more fundamental theory, a theory of "quantum gravity", which is still being worked on. However, this does not make Einstein's theory wrong. Theories, within certain limits that can be quantified, make good predictions and provide good explanations, in the sense discussed at the end of Chap. 4. Only QFT has so far turned out to be universal. Possibly quantum gravity, when found, will also be a variant of QFT. But perhaps at some point it will turn out that QFT itself goes back to something quite different by means of reduction by replacement.

Can we say that Newton's theory follows logically from Einstein's, as in reduction by generalization? That would be somewhat inaccurate, because in fact the predictions of Newton's theory diverge from those of Einstein's in many cases (namely, where we leave its range of validity). Moreover, Newton's theory says that gravity is a force field, and not only does this not follow from Einstein's theory, it actually contradicts it. So it is better to say: it follows from Einstein's theory why the world looks *as if* Newton's theory holds under certain conditions (namely, when the curvature of spacetime is sufficiently weak). From quantum gravity, in turn, it must follow why the world under certain conditions looks as *if* Einstein's theory holds.

Since neither Newton's nor Einstein's gravity has absolute validity and quantum gravity is still waiting to be discovered (and is possibly not yet the last word), it would be presumptuous to say that Einstein is right and Newton is wrong. Rather, every physical theory should be regarded as such an *as if.* We should never assume that we have grasped the fundamental units of reality with a theory. Everything may turn out to be something quite different at the next level. A theory is to be regarded as *correct* if it provides correct descriptions and predictions within its limits. Therefore one should also be very careful with ontological statements, i.e. statements about what the "really existing things" are. Since reduction by replacement occurs unexpectedly in the course of scientific progress (every such discovery is virtually tantamount to a scientific revolution, which is impossible to predict), things can always turn out to be an *as if in* retrospect.

The best known, but also most difficult reduction by replacement is the step from classical mechanics to QM. The quantum world is completely different from the world we think we know, and yet QM is a correct and confirmed theory that has more general validity than classical mechanics. Since the latter also originated with Newton, this was another reason to say he was wrong. The poor man didn't deserve that, because classical mechanics is also correct within the limits of its scope. From the point of view of QM, however, it turns out that the world only looks as *if* it behaves "classically" to us in the macroscopic realm. To deduce this *as if,* however, proves to be much more difficult in QM than in GR.

5.4   Reduction of Effective Theories

Another type of reduction with an as-if character is that which occurs in the formation of so-called **effective theories**. Let us take quantum chemistry as an example. If you want to derive the properties of atoms and molecules from

QM, it is sufficient to consider the electromagnetic interaction between positively charged atomic nuclei and negatively charged electrons (including the interactions of electrons with each other). The fact that the atomic nucleus itself consists of protons and neutrons is irrelevant for this task. From the point of view of chemistry, the atomic nucleus is an indissoluble entity. This is because in physics, roughly speaking, the energy required to resolve a structure (resolve in the microscope sense) is inversely proportional to the size of the structure. Since the nucleus is about 100,000 times smaller than the atom, the energy required to "see" the substructure of the nucleus is about 100,000 times greater than the energies involved in chemistry, specifically the energy of electrons. The electrons are completely "blind" to this substructure. Therefore, when you do chemistry, you can *pretend that* the nucleus of the atom is an elementary particle.

The QM of protons, neutrons and electrons is more fundamental than the QM of atomic nuclei and electrons. You get from the first to the second by a mathematical step called "integrating out degrees of freedom". The result is an effective theory.

In this case, reduction from the effective theory to the more fundamental theory is equivalent in content to reduction by decomposition. You get from the effective theory to the theory of protons and neutrons by establishing that the atomic nucleus has a substructure and can be split. But the crucial addition with the effective theory is the statement that this substructure does not matter for certain problems, e.g. chemistry, and that you can save the decomposition step in this case, and forget about the substructure. That is, the non-fundamental theory of atomic nuclei and electrons is fundamental enough for the problem (more precisely, for the chosen energy range) and gives accurate predictions. In this respect, reduction in effective theories can be seen as the exact opposite of reduction by decomposition. In the latter case, the decomposition is the crucial thing that yields the insight. In the former case, the insight is precisely that the decomposition is unnecessary.

An example of an effective theory that has nothing to do with decomposition comes from the so-called Kaluza-Klein theory . This is a theory with four spatial dimensions (i.e. five dimensions if you add time). The fourth dimension, unlike the other three, does not extend to infinity, but forms a circle with a tiny circumference, many orders of magnitude smaller than you can resolve with the best microscope in the world. So you can never observe this dimension, but live in an effective world with three spatial dimensions. What is the point of theoretically introducing a dimension if you can't observe it anyway? Well, the amazing insight of Kaluza and Klein was that with this step a unification of electromagnetism and gravitation can be achieved. If one endows the four-dimensional theory with Einsteinian gravity and then performs the "integrating out degrees of freedom" step along the fourth dimension, the result is an effective theory in three spatial dimensions in which Einsteinian gravity *and* Maxwellian electromagnetism are present. Electromagnetism thus turns out to be a side effect of gravity. A beautiful idea, but unfortunately it led to contradictions due to other theoretical considerations and thus had to be abandoned, at least in this simple form.

5.5 The Special Role of Physics

The hierarchy of the natural sciences is a hierarchy of fields, subfields and theories. The levels of this hierarchy are linked by the four forms of reduction mentioned above. In physics, all four forms occur equally. In the other fields, reduction by decomposition predominates, and has also led to most of the debate. That is why I have gone into it in particular detail. However, I find reduction by replacement particularly exciting because it changes the "overall character" of things. This plays a big role in the reasoning in the rest of the book.

Physics occupies a special place in this hierarchy, firstly because it lies at its lower end, and secondly because there is a greater variety of types of reduction within physics than elsewhere, as shown above. It follows from the first point that the fundamental limits of physics are also the fundamental limits of natural science. For if something can be expressed at all in terms of natural science, then in *principle* it can be expressed in terms *of* physics. But the practical limits of physics are *not* the practical limits of natural science. For since reduction by decomposition is only a reduction in *principle* and often cannot be carried out in practice, any field of natural science can arrive at findings whose derivation from physics is practically impossible.

Often the reduction in concrete cases does not bring any new understanding, is not interesting. It is enough to know that and how it is possible in *principle,* but to carry it out (e.g. to derive the structure of a certain molecule from QM) is not worth the effort. After all, each level has its own terminology, i.e., things and concepts are linguistically summarized there in a very specific way that facilitates working and understanding at that level. Jumping to a different level also means switching to a different terminology, with which one may be less familiar and also less able to efficiently describe what is going on at the higher level. This is also a consequence of weak emergence; the newness of the upper level demands new forms of expression.

Basically, the situation is similar to software development. Depending on the problem, a suitable higher programming language is used, which is expressed in the form of concepts addressing the mindset of the developer (programmer) and the requirements of the task, and summarizes things in such a way that efficient work is possible. However, the computer's processor only understands a small number of very specific instructions that are executed on small groups of bits (ones and zeros). So, in order to work, the developer's program must be translated into this **machine language**. This is done by a translation program called a **compiler**. This is also a form of reduction! The compiler turns a text file (or files) containing the developer's program code into an executable file containing machine language instructions. It is useful for the developer to know how the compiler works in *principle*, but only in rare cases will he bother to look at machine language code that is difficult for humans to read.

Because of the different terminologies of the different levels, each natural science has its own expressions and its own beauty. Reductionism *in principle does* not detract from the achievements and certainly not from the significance at the higher levels. But it leads to the fact that physics has a very special significance for philosophical questions, because it is the foundation of the hierarchy.

6

Physics

6.1 History and Overview

The name of physics comes from the Greek, but the real history of physics begins with the publication of Isaac Newton's masterpiece *Philosophiae Naturalis Principia Mathematica* (*The mathematical principles of natural philosophy*) in 1687. In this work, Newton defines the basic concepts of physics, describes the mathematical tools necessary to deal with these concepts, and establishes the first law describing one of the fundamental forces of nature: gravity.

Newton's work did not come out of nowhere; there were just in the seventeenth century a number of great scientists who prepared the way for it with their findings, including Descartes, Galilei, Kepler, Huygens, Torricelli. However, since the very decisive concept of a *force* is only defined in its final form by Newton, I would still count this (admittedly somewhat strictly) as part of the *prehistory of* physics.

The title of Newton's work also shows that physics is intimately connected with two other disciplines: mathematics and philosophy. The multifaceted relationships in this trinity will occupy us throughout the book.

Eighteenth-century physics was essentially concerned with drawing conclusions from Newton's work, with sifting through the terrain that Newton had staked out and that is now known as **classical mechanics**. New mathematical formalisms were established that allowed for more skillful calculations within the framework of Newton's theory. Practical applications were opened up (e.g., structural analysis, ballistics, oscillatory phenomena), and the theory, which at first applied only to idealized point-like objects and to rigid bodies,

© Springer-Verlag GmbH Germany, part of Springer Nature 2022
J.-M. Schwindt, *Universe Without Things*,
https://doi.org/10.1007/978-3-662-65426-2_6

was extended to fluids and gases (hydrodynamics). It was a very French century, and so the physics of this period is also dominated by French names: d'Alembert, Lagrange, Legendre, Laplace, Fourier, Coulomb. An exception were the Swiss Bernoulli and Euler, two main founders of hydrodynamics.

At the beginning of the nineteenth century, one then increasingly turned to new areas that lay outside Newton's terrain. On the one hand, these were the electrical and magnetic phenomena, which until then had represented a curiosity, of which it was not clear how it would ultimately fit into the picture. The focus of research was initially in France (Coulomb and Ampere, after whom the units for charge and current are still named today), but then increasingly shifted to Great Britain, where Faraday and Maxwell made the decisive steps towards a unified **electromagnetic theory**, which on top of that described light as an electromagnetic wave and thus also included the field of optics.

On the other hand, these were phenomena that had to do with heat and that played a role both in the construction of new machines, especially the steam engine, which was quite decisive for the beginning of industrialization, and in the course of chemical reactions, which were also intensively researched at that time. The result was **thermodynamics**, which at the end of the nineteenth century was developed into **statistical mechanics**, in which heat is understood as the statistical distribution of movements of small particles – atoms or molecules – of which matter is composed. Many researchers were involved in these developments, but the essential steps towards statistical mechanics were taken by Maxwell (Scotland), Boltzmann (Austria) and Gibbs (USA).

At the beginning of the twentieth century, things went from strength to strength. In 1905, Einstein established his **special theory of relativity** (SR), in which he combined space and time into a common geometric structure, *space-time,* thus explaining the curious observation that the speed of light is the same for every observer, no matter how fast he runs away from or towards a beam of light. SR also sheds a new light on electromagnetism, simplifying it. With its help, Maxwell's four equations can be reduced to a single equation. SRT is very simple, it can be easily formulated with middle school mathematics. On the other hand, it puts our intuition of space and time to a hard test, which is why many people find it difficult. Ten years later, Einstein followed up with the **general theory of relativity** (GR), in which he introduced a *curvature* of space-time and thus reduced the phenomenon of gravity to pure geometry.

Meanwhile, numerous experiments were conducted in connection with various forms of radiation and with atoms. This secured the status of atoms as the building blocks of matter and made many of their properties accessible. In particular, their internal structure of atomic nucleus and electron shell became clear. However, their behaviour could not be reconciled with Newtonian physics and the theory of electromagnetism. Instead, a whole new theory was needed to describe them, **quantum mechanics** (QM), which was established around 1926, with the main part of the formulation going back to Schrödinger (Austria), Heisenberg (Germany) and Dirac (England). QM is a very strange theory, it contradicts our intuition even more than Einstein's two theories of relativity. But it describes the behavior of atoms with incomparable precision and "explains" the entire periodic table of the elements and the occurrence of chemical reactions.

The experiments with radiation also revealed a previously unknown phenomenon, *radioactivity*. Over time, it became clear that this was due to processes in atomic nuclei, in particular those processes that split an atomic nucleus into two smaller ones. However, nothing was yet known about the harmful effects of radiation on the organism, which led to serious illnesses and the premature deaths of numerous researchers who had experimented with radioactive phenomena without concern. In the 1930s it was shown that the atomic nucleus consists of protons and neutrons and that their dynamics determine the behaviour of the nuclei and thus also of radioactivity. The knowledge gained was sufficient to develop the atomic bomb during the Second World War. However, the forces that determine the interaction of protons and neutrons, the so-called *weak* and *strong nuclear forces*, were only understood in more detail later, within the framework of particle physics.

Based on GR and measurements with ever better telescopes, theoretical work by the Russian Friedmann and the Belgian Lemaitre, together with observations by the American Hubble, led around 1929 to the discovery of the expansion of the universe and thus the *Big Bang model*. This laid the foundation for modern **cosmology**, i.e. the physics and history of the universe as a whole, and was quite crucial for our world view today.

The 1950s to 1970s were dominated by elementary particle physics. New particles were discovered all the time, which then had to be integrated into a common theory. The theoretical framework for this was **quantum field theory** (QFT), a further development of QM, which not only inherited all the strangeness of the latter, but was additionally plagued by nerve-wracking mathematical difficulties. Nevertheless, it describes phenomena of particle physics with unprecedented precision. It is by far the most universal theory we possess. Almost all of physics can be traced back to it, except for its application to gravity, which has yet to be clarified. Many physicists have been involved in its development, but the best known of them is probably the

American Feynman. In the course of the 1970s, the back and forth of newly discovered particles and new theories (all of them different variants of quantum field theories) finally converged towards one theory that encompasses all particles found up to that time and also all those found since: the **standard model of particle physics.**

The theories and fields mentioned in this overview also define the study of physics today. A typical course program includes four basic lectures in theoretical physics: (1) classical mechanics, (2) electromagnetism and special relativity, (3) quantum mechanics, and (4) thermodynamics or statistical mechanics. In the advanced studies (Master), (5) Quantum Field Theory (or Particle Physics) and (6) General Relativity are offered, as well as (7) Cosmology, whereby all other fields and theories play a role in the latter.

6.2   Physical Theories and Experiments

What is a physical theory? No one has answered this question more precisely and clearly than Pierre Duhem in his masterpiece *The Aim and Structure of Physical Theory* from 1906 (French: *La théorie physique, son objet et sa structure*); therefore we want to let him speak here and present his view. He defines:

> *A physical theory is a system of mathematical propositions deduced from a small number of principles, which aim to represent as simply, as completely, and as exactly as possible a set of experimental laws.* (Duhem 1954, p. 19)[1]

Thus, regularities that have been found experimentally are to be brought into a common context and presented as the logical consequence of a small number of "principles" (in mathematics one would say axioms) that define the specific theory. In establishing such a theory, according to Duhem, four fundamental operations are to be performed:

1. The definition and measure of physical quantities
2. The choice of hypotheses
3. The mathematical development of the theory
4. The comparison of theory with experiment

Regarding **point 1**: Theories refer to certain physical quantities. Thereby, a connection between mathematical symbols, with which the theory is formed,

---

[1] The year here refers to the English edition.

and the physical properties themselves is established by a system of measurements. Among these, a distinction must still be made between simple and composite properties:

> *Among the physical properties which we set ourselves to represent, we select those we regard as simple properties, so that the others will supposedly be groupings or combinations of them. We make them correspond to a certain group of mathematical symbols, numbers, and magnitudes, through appropriate methods of measurement. These mathematical symbols have no connection of an intrinsic nature with the properties they represent; they bear to the latter only the relation of sign to thing signified.* (Duhem 1954, p. 20)

In classical mechanics, for example, length, time and mass are the "simple" properties. All other properties are composed of these; for example, a velocity is equal to the length of a distance traveled divided by the time it took to travel it. Length can be measured by a tape measure, time by a clock, mass by a scale (for simplicity, let's ignore the difficulty that a scale usually measures weight, which is not the same as mass). Through these methods of measurement, properties are assigned numerical values, which are given specific units. Lengths are measured in meters, time in seconds, masses in kilograms, velocities accordingly in meters per second, and so on. In thermodynamics, temperature plays a fundamental role. It is measured with a thermometer and expressed in degrees. In this way, all properties that we initially perceive sensually and qualitatively (something feels warm or cold) are translated into quantities with which a theory can then work:

> *The physical universe, which our senses present to us as an immense assemblage of qualities, had therefore to be offered to the mind as a system of quantities.* (Duhem 1954, p. 113)

In theory, these quantities are expressed by certain symbols, for example l for length, t for time, m for mass, T for temperature, and then, in order to *apply* the theory to a concrete situation, concrete numerical values must be used for these symbols.

Regarding **point 2**, the choice of hypotheses: Some principles are to be chosen (we can also call them basic assumptions, axioms, or postulates; they can also be equations that form the basis of a theory) that describe hypothetical relationships between physical quantities. In classical mechanics, for example, Newton's third postulate would be such a principle: *actio* equals *reactio*. This means that if a body A exerts a force on a body B, then B also exerts an equally strong force on A, only in the opposite direction. This principle can be seen, for example, in the recoil of firing a firearm: the recoil we experience is just as strong as the force that hurls the bullet forward out of the barrel (only the bullet is accelerated much more because of its small mass, because acceleration equals force divided by mass). The principle is also evident in gravity (gravitation): the earth attracts the sun just as strongly as, conversely, the sun attracts the earth. Just because the sun is about 300,000 times heavier than the earth, this is hardly reflected in any significant movement of the sun; the earth revolves around the sun, not vice versa.

In **point 3**, logical conclusions are then to be derived from the hypotheses, usually in the form of equations that relate the physical quantities to each other. These derivations are purely mathematical. The hypotheses together with the totality of all possible conclusions (in principle, there are always an infinite number of them!) then form the theory.

Finally, **point 4**, the comparison with experiment, is crucial. It marks the most essential difference between physics and mathematics. For the first three points are virtually identical to the formation of theory in pure mathematics: firstly, the choice of basic concepts, secondly, the choice of axioms that define relationships between the basic concepts, thirdly, the logical derivation of theorems. In physics, however, the first point also includes the description of measurement methods, which make the fourth point – comparison with experiment – possible in the first place by defining the connection between the mathematical quantity and the observation in the experiment. Comparison with experiment is also the only criterion of truth of a theory:

> *The aim of all physical theory is the representation of experimental laws. The words "truth" and "certainty" have only one signification with respect to such a theory; they express concordance between the conclusions of the theory and the rules established by the observers.* (Duhem 1954, p. 144)

The comparison with experiment, however, is often much more difficult than it first appears. As already mentioned, a theory contains an infinite number of conclusions which, in principle, could be tested experimentally. On the other hand, there are also an infinite number of conceivable experimental constellations to which the theory could be applied. It is therefore necessary to make an appropriate selection.

There is also the problem of translation between experiment and theory. The starting point for this is, as stated above, the specification of the measurement methods in point 1:

In order to introduce the circumstances of an experiment into the calculations, we must make a version which replaces the language of concrete observation by the language of numbers; in order to verify the result that a theory predicts for that experiment, a translation exercise must transform a numerical value into a reading formulated in experimental language. As we have already indicated, the method of measurement is the dictionary which makes possible the rendering of these two translations in either direction. (Duhem 1954, p. 133)

However, this is not so trivial:

But translation is treacherous: traduttore, traditore (to translate is to betray). There is never a complete equivalence between the two texts when one is a translated version of the other. Between the concrete facts, as the physicist observes them, and the numerical symbols by which these facts are represented in the calculations of the theorist, there is an extremely great difference. (Duhem 1954, p. 133)

The problem is that most measurement methods are quite indirect and in turn already *require* quite a few physical laws to be valid. A clock, for example, works on the basis of certain physical laws. In the case of an old-fashioned pendulum clock, this was the oscillating behavior of a pendulum. In a quartz clock, a much more complicated effect comes into play, namely "electromechanical resonance oscillations" of small quartz crystals. Modern atomic clocks are even more complicated, but also function more precisely. The physical laws governing the mechanisms of the respective clocks must already be assumed by a physicist so that he can assume that the digits displayed by the clock actually denote seconds, and so that he can use the clock to test *other* physical laws in an experiment.

If he measures the mass of an object with a scale, he must assume the mechanism of the scale to be correct. Strictly speaking, this measures a weight force exerted on it, i.e., the physicist must first accept the law of gravity and know the strength of the gravitational field on the earth's surface in order to accept how the scale translates a weight force into a mass. Further, he must accept the inner mechanism of the scale, that is, how it translates the strength of the weight force into the display of a number. In old-fashioned spring scales, this was done by a helical spring that was pulled apart in proportion to the force; in modern electromechanical scales, several electrical and mechanical effects are combined.

If he measures the temperature by means of an old-fashioned mercury thermometer, he actually reads the height of the mercury column. He must *presuppose* how this height is related to temperature, namely because of the expansion properties of mercury. In his experimental protocol, however, he will only record the measured temperature. He will not write anything about the height of a column of mercury and the theoretical reasons why he interprets this as a temperature measurement.

An experiment in physics is the precise observation of phenomena accompanied by an INTERPRETATION of these phenomena; this interpretation substitutes for the concrete data really gathered by observation abstract and symbolic representations which correspond to them by virtue of the theories admitted by the observer. [...]

The result of the operations in which an experimental physicist is engaged is by no means the perception of a group of concrete facts; it is the formulation of a judgment interrelating certain abstract and symbolic ideas which theories alone correlate with the facts really observed. [...]

Not one of the words serving to state the result of such an experiment directly represents a visible and tangible object; each of them has an abstract and symbolic meaning related to concrete realities only by long and complicated theoretical intermediaries. [...]

Between an abstract symbol and a concrete fact there may be a correspondence, but there cannot be complete parity; the abstract symbol cannot be the adequate representation of the concrete fact, the concrete fact cannot be the exact realization of the abstract symbol. [...] Between the phenomena really observed in the course of an experiment and the result formulated by the physicist, there is interpolated a very complex intellectual elaboration which substitutes for the recital of concrete facts an abstract and symbolic judgment. (Duhem 1954, p. 147 ff.)

Thus, in the choice of measurement methods from step 1 in the establishment of a new theory, several *other* theories are usually already "hidden", which must be assumed to be valid in order for the method to be accepted. In this way, physical theories build on each other, are interrelated, and the whole of physics takes on a holistic character as a result. This also applies to the concepts with which physics deals. Only the simplest ones arise directly from our sensory experience, such as length or time. Already the concept of mass contains some abstraction, even more the concept of energy. These terms are used because they have turned out to be useful in the interplay of experiment and theory. They also build on each other. Electromagnetism includes terms such as electric charge, current, voltage, electric field strength and magnetic field strength. These have emerged from a centuries-long process of abstraction, from a huge number of observations, interrogations of nature in experiments and theoretical considerations about them. For later theories, however, they are presupposed, serving as a basis for even deeper abstractions, such as in atomic physics:

The symbolic terms connected by a law of physics are, on the other hand, not the sort of abstractions that emerge spontaneously from concrete reality; they are abstractions produced by slow, complicated, and conscious work, i.e., the secular labor which has elaborated physical theories. If we have not done this work or if we do not know physical theories, we cannot understand the law or apply it. [...]

A physical law is a symbolic relation whose application to concrete reality requires that a whole group of laws be known and accepted. [...]

Physical science is a system that must be taken as a whole; it is an organism in which one part cannot be made to function except when the parts that are most remote from it are called into play, some more so than others, but all to some degree. [...]

If the interpretation of the slightest experiment in physics presupposes the use of a whole set of theories, and if the very description of this experiment requires a great many abstract symbolic expressions whose meaning and correspondence with the facts are indicated only by theories, it will indeed be necessary for the physicist to decide to develop a long chain of hypotheses and deductions before trying the slightest comparison between the theoretical structure and the concrete reality. (Duhem 1954, pp. 167 ff, 187, 204)

This shows that physics is a very laborious process. But it also dispels the illusion that one can understand physics virtually on the fly, summarize or explain one of the advanced theories in a few sentences. Learning physics takes time, a lot of time.

The holistic character of physics means in reverse: If now a discrepancy is found between theory and experiment, it is not clear at first whether the concrete hypothesis which the experiment means to test was itself thereby disproved, or whether one of the numerous presuppositions which entered into the interpretation of the experiment was wrong:

An Experiment in Physics Can Never Condemn an Isolated hypothesis but only a whole theoretical group. (Duhem 1954, p. 183)

Therefore, in order to check where exactly the deviation lies, one has to use many other experiments for comparison, which make use of the different presuppositions and the different conclusions of the theory in different ways:

The only experimental check on a physical theory which is not illogical consists in comparing the entire system of the physical theory with the whole group of experimental laws, and in judging whether the latter is represented by the former in a satisfactory manner. (Duhem 1954, p. 200)

The Approximate Nature of the Theories

Another important aspect of the comparison between theory and experiment is the handling of measurement inaccuracies. For on the theory side, concrete numerical values are inserted into the equations, but on the experiment side, one is always dealing with observations and measurements that exhibit small inaccuracies. The size of these inaccuracies has to be estimated by the experimenter. This is then often implicitly expressed by rounding to a certain number of decimal places. For example, if my scale tells me that I weigh 81.7 kg, then this is to be interpreted as my weight being between 81.65 and 81.75 kg; the inaccuracy is therefore about 0.1 kg. On the theory side, it is then again necessary to estimate how the inaccuracy continues in the equations.

Suppose we want to check an equation that occurs in the context of a theory. The equation links several physical quantities together. In that case, in a typical theory-experiment comparison, we fix all but one of the values and use the equation to make a prediction for the remaining value, which we can then in turn compare to the experiment. But if the given values are subject to inaccuracies of a certain magnitude (due to the inaccuracy of their measurement), what is the *theoretical* inaccuracy of the prediction for the remaining value, and how does this compare with the *experimental* inaccuracy in the measurement of that very value? Are theory and experiment in harmony with each other, given all these inaccuracies? These are the questions that the experimentalist has to ask himself.

In other cases, a differential equation is to be tested which describes the temporal development of a physical quantity. In this case, for example, the numerical value of the quantity is measured at two different times. The earlier value is used as the initial value in the equation and a prediction for the later value is derived from it, which in turn is compared with the measured later value. But if the initial value has an inaccuracy of the and the quantity, how does that affect the inaccuracy of the prediction?

Here there are "benign" and "malignant" cases. In the benign case, the inaccuracy of the prediction is at most as large as that of the initial value used. In the malignant case, a very small inaccuracy in the input value results in a very large inaccuracy in the prediction. For example, if I release a billiard ball at the top of a hill, millimeters in the initial position can determine which way the ball rolls down the hill.

Such considerations determine how well a comparison between theory and experiment can be made in a given case. It may also happen that a theory is initially confirmed by experiment, but is then found to be invalid when new experimental methods make higher accuracy possible. However, since complete accuracy can never be established, any theory must be regarded as an **approximation**, valid within certain limits of accuracy.

Moreover, theories are usually restricted to certain **scales**. That is, they hold with acceptable accuracy if the physical quantities they deal with are themselves within a certain range of values, within certain orders of magnitude. Newtonian mechanics applies with high accuracy as long as the velocities of all objects involved in an experiment are significantly less than the speed of light. As one approaches the speed of light, Newtonian mechanics is no longer sufficient; one needs SR. It loses its validity also on very small *length scales*, in the atomic region, which is governed by QM. In particle physics, one observes that the behavior changes depending on the energies with which particles are collided. So here the energy scale is crucial. For example, below a certain energy, the electromagnetic force and the weak nuclear force appear as two different forces with different properties. Above this energy, however, the two forces combine to form a single force.

Thus, progress in physics comes about, among other things, as greater accuracies or previously unexplored scales come within reach of experiments, for example, higher energies in particle accelerators. In the course of this progress, it also happens that new terms have to be introduced and the old ones are restricted in their scope of validity:

Physical law is provisional not only because it is approximate, but also because it is symbolic: there are always cases in which the symbols related by a law are no longer capable of representing reality in a satisfactory manner. [...]

Not that we must understand this to mean that a physical law is true for a certain time and then false, but at no time is it either true or false. It is provisional because it represents the facts to which it applies with an approximation that physicists today judge to be sufficient but will some day cease to judge satisfactory. Such a law is always relative; not because it is true for one physicist and false for another, but because the approximation it involves suffices for the use the first physicist wishes to make of it and does not suffice for the use the second wishes to make of it. [...]

The mathematical symbol forged by theory applies to reality as armor to the body of a knight clad in iron: the more complicated the armor, the more supple will the rigid metal seem to be; the multiplication of the pieces that are overlaid like shells assures more perfect contact between the steel and the limbs it protects; but no matter how numerous the fragments composing it, the armor will never be exactly wedded to the human body being modelled. [...]

To any law that physics formulates, reality will oppose sooner or later the harsh refutation of a fact, but indefatigable physics will touch up, modify, and complicate the refuted law in order to replace it with a more comprehensive law in which the exception raised by the experiment will have found its rule in turn. Physics makes progress through this unceasing struggle and the work of continually supplementing laws in order to include the exceptions. [...]

Physics does not progress as does geometry, which adds new final and indisputable propositions to the final and indisputable propositions it already possessed; physics makes progress because experiment constantly causes new disagreements to break out between laws and facts, and because physicists constantly touch up and modify laws in order that they may more faithfully represent facts. [...]

Physics [...] is a symbolic painting in which continual retouching gives greater comprehensiveness and unity, and the whole of which gives a picture resembling more and more the whole of the experimental facts, whereas each detail of this picture cut off and isolated from the whole loses all meaning and no longer represents anything. (Duhem 1954, pp. 172 ff, 204)

Many physicists, however, hope that one day a theory will be found which is a perfect "suit of armour" after all – a theory which represents *all the* facts and is no longer refuted by any experiment (discussion in Chaps. 8 and 9). Apart from this hope, the point that the theories of physics are approximations and provisional is quite crucial to the proper interpretation of physical discoveries. As already discussed in Chap. 5, both Newtonian and Einsteinian theories represent approximations; they apply within certain limits of precision and within certain *scales*. The range over which Einstein's theory can be applied is wider than that of Newton's theory. But this does not mean that one theory is "right" and the other is "wrong".

Physics and Metaphysics

A controversial point in Duhem's work is his rejection of the idea that physical theories *explain* anything. He justifies this by saying that physics must distinguish itself from metaphysics:

While we regard a physical theory as a hypothetical explanation of material reality, we make it dependent on metaphysics. In that way, far from giving it a form to which the greatest number of minds can give their assent, we limit its acceptance to those who acknowledge the philosophy it insists on. (Duhem 1954, p. 19)

To illustrate this, he describes how different philosophical schools (in a time before Maxwell's theory of electromagnetism was established) argued about how a magnetic field could come about. Each of these schools had its own preconceptions about how reality should be:

- From the point of view of the *peripatetics*, everything is substantially composed of *substance* and *form*, whereby "form" here is a very broad term that can encompass all conceivable properties, both perceptible and hidden, and thus in particular also something like a "magnetic" property.
- From the point of view of *Newton's* followers, however, all matter had to consist of point-like masses that exert certain forces on each other, even over a distance (action at a distance, as in gravitation). These forces were to describe by laws of nature. Additional properties of the mass points were not allowed.
- For the *atomists*, however, remote effects were pure illusion. Matter had to consist of small hard objects, the atoms, which had a certain mass and a certain shape (especially not simply point-like). These atoms could only act on each other by colliding. Over a distance, they could not affect each other in any way.
- The *Cartesians* also did not believe in action at a distance. In their view, all space was filled with a homogeneous, incompressible "fluidum" whose vortical motions created the illusion of matter in the conventional sense.

Each of these schools accordingly attempted to construct theories of magnetism that were consistent with their prejudices. None of these theories got very far; in particular, each found adherents only within its own philosophical camp. Therefore, Duhem concludes, one must keep such metaphysical prejudices out of physics from the start. One must understand each theory only as what it is, namely a system of mathematical theorems for the compact description of experimentally found regularities, without making any additional assumptions about the nature of reality. The price one pays for this is the abandonment of theory as an *explanation of* anything. For if the mathematical theorems are no longer linked to a statement about the nature of reality, if they are only purely mathematical statements that refer to experimental results in a sometimes complicated way, then they lose all that we expect from an explanation.

Today, this can be observed very well in the example of QM. This theory describes the behaviour of atoms almost perfectly. However, it contradicts all our intuitions about the nature of matter (Sect. 7.6). Nobody who learns QM has the feeling that anything is really explained in a deeper way, except in a purely mathematical sense: For example, we understand *how* the equations of QM lead to the structures of the periodic table of the elements, in so far it "explains" these structures. But how a reality can be such that it makes possible such an "impossible" theory as QM remains unclear. Here, new philosophical schools have emerged that *interpret* QM in different ways, attempting to grasp its connection with reality in different ways. But the crucial thing is precisely that QM as a physical theory functions completely independently of these interpretations, as a physical theory that makes correct predictions for experiments without having to *explain* anything.

However, the dissatisfaction of many physicists with QM also shows why these theses of Duhem are so controversial. We just wish for an explanation by a theory. Without the hope for explanations, a very important motivation to do physics would fall away. And in the *classical* theories, i.e. the theories before QM, we do see such explanations. In SR, for example, we have the feeling that it tells us something real about space and time, and that something is *explained by* it, for example the constancy of the speed of light. We also have the strong feeling that physics could not work so well and could not summarize so many phenomena in so few laws if these laws did not reflect reality itself.

However, Duhem also sees this, but he sees it for what it is: a feeling, a hunch, a belief. From this he develops the concept of *natural classification* and once again separates this from the concept of *explanation*:

> Without claiming to explain the reality hiding under the phenomena whose laws we group, we feel that the groupings established by our theory correspond to real affinities among the things themselves. [...]
>
> Thus, physical theory never gives us the explanation of experimental laws; it never reveals realities hiding under the sensible appearances; but the more complete it becomes, the more we apprehend that the logical order in which theory orders experimental laws is the reflection of an ontological order. [...]
>
> Thus the analysis of the methods by which physical theories are constructed proves to us with complete evidence that these theories cannot be offered as explanations of experimental laws; and, on the other hand, an act of faith, as incapable of being justified by this analysis as of being frustrated by it, assures us that these theories are not a purely artificial system, but a natural classification. [...]
>
> This characteristic of natural classification is marked, above all, by the fruitfulness of the theory which anticipates experimental laws not yet observed, and promotes their discovery. [...]
>
> What is lasting and fruitful in [most physical doctrines] is the logical work through which they have succeeded in classifying naturally a great number of laws by deducing them from a few principles; what is perishable and sterile is the labor undertaken to explain these principles in order to attach them to assumptions concerning the realities hiding underneath sensible appearances. (Duhem 1954, pp. 26 ff, 38)

In the words of Karl Jaspers, one might say: that physical theories reflect a deeper reality is a truth we physicists live by, not one we can prove. And this truth has suffered some mortification by QM.

So metaphysics is to be kept out of physical theories. What about the reverse case? Should the results of physics be used to discuss metaphysical or, more generally, philosophical concerns? Duhem does not comment on this, but looking at the matter, it is clear that the answer must be yes. While physics cannot determine what the underlying reality is, it can make strong negative statements. It can rule out metaphysical hypotheses that conflict with the laws of physics. For while the theories of physics are only approximations and provisionalities, they are nevertheless often well **confirmed** within their range of validity, i.e., they describe the observed phenomena in that range to a very good approximation, and so can say a great deal about how things definitely do *not* behave. This alone can be used to refute all four of the schools just discussed: Atoms are neither hard spheres that interact only via collisions, nor are they point-like masses in the Newtonian sense. The peripatetic concept of matter and form is not tenable as it stands, and the Cartesian fluid is also not consistent with the known physical laws.

Astronomy and cosmology have come to exclude worldviews with an eternally constant cosmos, just as they have long excluded worldviews in which the earth is an absolute center. In fact, physics has a huge range of statements of philosophical consequence to make. It provides numerous viewpoints on space, time, the nature of matter, the history of the universe, the origin of the earth, the sun, and the moon, determinism and chance, cause and effect. It opens up new perspectives on these topics, with a complexity and so full of surprising aspects that philosophers would never have thought of them on their own. In doing so, it poses great challenges to metaphysics: Reality must be such that it is consistent with the known laws of nature, even QM. For this is the way it is: physics has to make demands on metaphysics, not vice versa.

Thus, while physics cannot answer the deep questions about the nature of reality, it has a wide and inspiring range of viewpoints to contribute that take our not knowing to a much higher level.

Therefore a physicist should clearly separate physics and metaphysics. But he should not therefore renounce philosophizing. Because physics is so difficult and time-consuming, it needs the physicists themselves to recognize the philosophically relevant content of a theory and to think it further. This is hardly possible for an outside pure philosopher. Nothing else is done by Duhem, who in his work virtually designs a philosophy of physics.

Beauty and Abstraction

We will briefly touch on three other aspects of physics that Duhem highlights in his book. First, there is the *beauty* expressed in the theories:

> *Order, wherever it reigns, brings beauty with it. Theory not only renders the group of physical laws it represents easier to handle, more convenient, and more useful, but also more beautiful. It is impossible to follow the march of one of the great theories of physics, to see it unroll majestically its regular deductions starting from initial hypotheses, to see its consequences represent a multitude of experimental laws down to the smallest detail, without being charmed by the beauty of such a construction, without feeling keenly that such a creation of the human mind is truly a work of art.*
> (Duhem 1954, p. 24)

Indeed, this is something that every physicist will probably feel at one point or another when studying the subject. The beauty that unfolds is one of the greatest rewards for those who devote themselves to physics. Caution, on the other hand, should be exercised in drawing the reverse conclusion that a hypothesis which is felt to be beautiful must also find its way into a valid theory. This temptation has led many a theorist astray.

Secondly, it should be mentioned that Duhem contrasts the ability to abstract with the ability to imagine, treating them as opposites, clearly preferring the former to the latter. Accordingly, he is highly critical of representations of physics that focus on vividness. In a chapter that seems somewhat bizarre today, he contrasts the Franco-German approach with the English, which he considers inferior (excluding the great Newton, of course). In fact, at that time it was fashionable in English physics (i.e. at English universities and in English textbooks) to illustrate all physics, especially electrodynamics, by mechanical models and to clothe them in corresponding pictures:

> *The French or German physicist conceives, in the space separating two conductors, abstract lines of force having no thickness or real existence; the English physicist materializes these lines and thickens them to the dimensions of a tube which he will fill with vulcanized rubber. In place of a family of lines of ideal forces, conceivable only by reason, he will have a bundle of elastic strings, visible and tangible, firmly glued at both ends to the surfaces of the two conductors, and, when stretched, trying both to contract and to expand. [...]*
>
> *Here is a book intended to expound the modern theories of electricity and to expound a new theory. In it there are nothing but strings which move around pulleys, which roll around drums, which go through pearl beads, which carry weights; and tubes which pump water while others swell and contract; toothed wheels which are geared to one another and engage hooks. We thought we were entering the tranquil and neatly ordered abode of reason, but we find ourselves in a factory. [...]*

No doubt, wherever mechanical theories have been planted and cultivated, they have owed their birth and progress to a lapse in the faculty of abstracting, that is, to a victory of imagination over reason. [...]

A gallery of paintings is not a chain of syllogisms (Duhem 1954, p. 70 ff., 86)

Duhem warns of the negative consequences should such an approach also become established in French schools and universities. Physics students should be trained in abstraction, not lulled by vivid but ultimately false images. What can we learn from this discussion today? Representations of electricity theory using factory-like pictures have disappeared from textbooks. The study of theoretical physics is very abstract. Today, one can rarely complain about too much vividness. In popular scientific presentations, however, one tries more than ever to be vivid and to clothe highly complicated relationships, which can only be expressed in the language of higher mathematics, in pictures understandable to the layman. This can sometimes be quite suitable for conveying an initial idea of what a theory is actually about. Nevertheless, caution is required here. These pictures usually distort the actual contents of a theory very strongly, tear parts out of a larger context, and can also lead to misunderstandings and misinterpretations. A layman might get the impression: finally someone did explain that stuff well, finally he did understand that stuff. But he should interpret this correctly. He has only received a first impression, in many cases a distorted picture. Under no circumstances can a few sentences and pictures really summarize the content of a theory, under no circumstances can they replace a study of physics, which fully describes the theory and the terms it uses, as well as the mathematical language it uses, and places it in the overall context of physics, where, as we have learned, theories build on each other and form a holistic whole.

Finally, Duhem emphasizes that the whole edifice of physics and natural science as a whole must be based not on logic alone but also on *common sense,* again the kind of truth we live by but cannot prove:

In this situation, as in all others, science would be impotent to establish the legitimacy of the principles themselves which outline its methods and guide its researches, were it not to go back to common sense. At the bottom of our most clearly formulated and most rigorously deduced doctrines we always find again that confused collection of tendencies, aspirations, and intuitions. No analysis is penetrating enough to separate them or to decompose them into simpler elements. No language is precise enough and flexible enough to define and formulate them; and yet, the truths which this common sense reveals are so clear and so certain that we cannot either mistake them or cast doubt on them; furthermore, all scientific clarity and certainty are a reflection of the clarity and an extension of the certainty of these common-sense truths. (Duhem 1954, p. 104)

6.3   Physics and Mathematics

Physics and mathematics are closely linked. Mathematics is the language in which physical theories are written. The two disciplines have always cross-fertilized each other. The development of infinitesimal calculus by Newton and Leibniz made classical mechanics possible in the first place. New mathematical insights, structures and methods often found applications in physical problems only shortly thereafter. Conversely, physics regularly challenged mathematics, demanded extensions of existing formalisms, new methods, inspired it by its questions. Sometimes physicists already intuitively used new mathematical concepts for "calculation tricks", which only afterwards were substantiated and justified by mathematicians by precise definitions and a corresponding theory (an example is the theory of distributions).

Sometimes physicists have an intuitively better understanding of a mathematical problem than mathematicians themselves, because they understand the problem in the context of a physical system whose behavior they can well imagine from their experience. For example, the physicist Edward Witten received the Fields Medal, the highest award for mathematicians.

We can ask ourselves: Why is the world actually so mathematical, why is mathematics the *right* language to formulate physical theories, why is physics so successful with it? Does it have to be that way? Is the world intrinsically mathematical? Should that surprise us? Or is it something man-made? Are we imposing mathematics on the world? Are we pruning it with mathematics?

These are difficult questions. Like most physicists, I believe that with physics we achieve a *natural classification* in Duhem's sense. The whole thing just works far too well to be just something imposed, superimposed. The mathematizability of large parts of our realm of experience is an essential aspect that comes from reality itself, not something man-made (this is a belief, not something I can prove). But the question remains whether *everything* about the world at a fundamental level is mathematical and *only* mathematical, as Max Tegmark, for example, claims, or whether there is something else besides. I would like to postpone this question to Chap. 11.

I think that the mathematizability of large parts of our field of experience is a basic prerequisite for us to be able to experience a world as a *common world at* all. The common comes about by the fact that we can communicate with each other and thereby, at least in many cases, *agree* on something, at least recognize a few *facts* together. If communication always had to remain poetically fuzzy, simply because the world was poetically fuzzy and not mathematically exact, we would not get very far with facts, agreement and common ground. It is only because we can agree on *precise* facts – length, width, height, weight, the number of things, geometric shapes, date, time, and so on – that communication reaches a level that can be called an *exchange of information*. *In* my opinion, these precise facts are only possible in a world that contains some basic mathematical properties. An advanced project like natural science, which is based on a very complex network of very precise facts, can only work, I think, if mathematizability goes very far. So I think that, in principle, natural science can only function in a world that has a highly mathematical character.

The approximate nature of physical theories makes that physicists deal with mathematics differently from mathematicians in many respects. For the mathematician, there is no such thing as an approximation; he wants to prove and calculate things exactly, without any doubt and without any inaccuracy. The physicist, on the other hand, wants only to estimate. The calculations he has to perform within the framework of a theory are often very complicated. Therefore, he estimates which expressions in his calculation will make only such small contributions that they are below the accuracy limit relevant to the problem he is working on, and will simply ignore these expressions and confine himself to the essentials.

The approximate character also implies that some things which the mathematician regards as a major problem leave the physicist completely cold. For example, in the theory of fluids it is well known that under certain conditions they behave according to certain differential equations, the Navier-Stokes equations. But it is an unsolved mathematical problem whether these equations have solutions at all under certain conditions. In fact, it is one of the seven Millennium Problems, for the solution of which a prize of one million dollars is offered. Many a mathematician thinks that it must worry a physicist that an equation that appears in his theories may not be solvable at all. But the physicist remains quite calm. For the mathematician naturally means an *exact* solution. The physicist, however, knows that the Navier-Stokes equation in his case is only an approximation. He even knows at which scale it loses its validity (at the latest at the atomic scale, where matter is "granular" and no longer has the property of a uniform fluid). He is not looking for exact solutions at all. He knows the range in which the equations are a good approximation and is able to determine approximate solutions in this range and to describe their qualitative behaviour.

So in many cases it is justifiable for the physicist to handle the mathematics in a way that the mathematician finds rash or imprecise. But there are other cases in which physicists do indeed sloppily handle mathematics in a way that cannot be justified even by the approximate nature of the theories. This sometimes happens in order to produce numerical results more quickly "on spec", which can then be compared with an experiment and, if successful, published. A particularly serious case was the handling of quantum field theory (QFT) in its early years. Using the correct mathematics, the formalism used at that time produced "infinity" in almost every calculation. Because one could not do anything with it, one used all kinds of tricks to get rid of these infinities. These tricks were chosen quite arbitrarily and had not much to do with clean mathematics. Strangely enough, however, they produced results that agreed with the experiment. That is why the method prevailed. This was another meaning of the "shut up and calculate" mentality mentioned in Chap. 2. Now this was not about ignoring philosophy, but ignoring mathematics. Basically wrong calculations were to be carried out because they led to the right results. Only later, in the theory of the so-called *renormalization group,* which describes an essential aspect of QFT (Sect. 7.7), the matter was cleared up and the results could be understood with *correct* mathematics.

7

The Pillars of Physics

In the following, we will take a closer look at the *confirmed* theories of physics, i.e. those theories whose validity within a certain area – their respective area of competence – has been confirmed and secured by a wealth of experiments. It will be shown what these theories are about in each case, which terms they use, and what they have to contribute to our view of the world.

Much of it – especially in the later parts – sounds admittedly implausible, and we physicists would not believe it ourselves if nature did not force us to. Since this is a popular science book, I have to refrain from using higher mathematics, i.e. the language in which these theories are formulated. A translation into the terms of everyday language, in the form of metaphors and explanations, is possible to a certain extent, but only to a certain extent. At the same time, such representations are always prone to misunderstanding. I already try to avoid some of the quite common misunderstandings I have observed in readers of popular science literature, but this can never be completely successful.

This chapter can therefore only be, as announced in Chap. 1, a "treasure map", providing the reader with an overview and some suggestions. It is also intended to encourage the mathematically gifted among the readers not to be satisfied with these explanations, but to penetrate into the spheres of higher mathematics themselves and to study the theories in their natural language.

Only in the case of SR do I take the liberty of demonstrating a little of the actual mathematical basis of the theory. This is possible because this theory is mathematically the simplest of all; it can be mastered with middle school mathematics. If this is already too much for you (some people feel a strong

© Springer-Verlag GmbH Germany, part of Springer Nature 2022
J.-M. Schwindt, *Universe Without Things*,
https://doi.org/10.1007/978-3-662-65426-2_7

aversion to anything that throws them back to their experiences with school mathematics), you may safely skip the relevant section. This of course also applies to all other sections, where it gets "too much" for you.

7.1   Classical Mechanics

As discussed at the end of Chap. 3, the development of the *differential calculus* was crucial to the birth of modern physics. Newton was able to work with its concepts when he published *Philosophiae Naturalis Principia Mathematica* in 1687, the most important book in the history of physics (often abbreviated simply to *Principia*; every physicist immediately knows what is meant). The title says it all: natural philosophy is founded on mathematical principles. That is exactly what physics is. The work is crammed with imaginative geometrical constructions. Geometry was still considered the supreme discipline in mathematics at the time, especially after the work of René Descartes a few decades earlier. In more modern accounts, geometry is used much less, and algebra is used almost exclusively. In differential calculus, Leibniz's notation has prevailed, not Newton's. For these reasons, and because it was written in Latin (as almost all scientific works were at the time), the work is difficult to read for a physicist of today. Nevertheless, it is the starting point and the cornerstone of all physics.

Forces

The central insight expressed in the *Principia* is that **force** is the central concept for understanding nature. The laws of physics are laws of forces. Newton transformed the colloquial concept of force into a precise mathematical definition: Force is mass times acceleration. (This means, for example: If your car weighs twice as much as your neighbor's, your engine must exert twice the force to achieve the same acceleration).

Another realization is that you can combine forces with each other: When you hold a dumbbell by your outstretched arm, gravity pulls it downward. You have to exert the same force in the opposite direction to keep the dumbbell in position, i.e. to prevent acceleration.

In physical problems, the forces are often determined by a given situation. The forces can either be constant or depend on the position or time. Division by the mass gives the resulting acceleration, and from this the whole course of motion can be determined by differential calculus.

The most general law of force Newton found was that of **gravitation**. It makes things fall to the ground, it keeps the moon in its orbit around the earth, the earth and the other planets in their orbits around the sun, and the sun in its orbit around the center of the Milky Way (Newton did not yet know about the latter). The law of gravitation states that two objects always attract each other, with a force proportional to the two masses, and inversely proportional to the square of the distance. (For example, at twice the distance, the force is four times less).

A typical physics problem might now read, "Given a planet orbiting a sun with the following mass, where the planet has the following position and velocity relative to the sun at the initial time t_0. Calculate the exact trajectory." As a result, we get that the trajectory is an ellipse whose parameters are determined by the initial conditions.

In addition to gravity, three other basic forces are known today. The best known of these is the electromagnetic force, which occurs as electricity or magnetism, or a combination of both, depending on the situation. In addition, there are two "nuclear forces"- a "strong"and a "weak"one – which get their name from the fact that their range is very short, about as short as an atomic nucleus is large. The strong nuclear force holds the atomic nucleus together (which would otherwise be torn apart due to its positive charge), the weak one is responsible for various radioactive reactions.

All the forces we know today go back to these basic forces. For example, when two objects collide (billiard balls, for example, or my head with the wall), why don't they just pass through each other, like the ghosts in some scary movies, but exert such a strong force on each other? (You can see that the force is very strong because the change in velocity only occurs in the very brief instant of collision, so the acceleration, i.e. the change in velocity per time, is very large). Why doesn't my cup just fall through the table top? Actually, the electron shells of the outermost atomic layers of the two objects do penetrate each other a tiny bit, but in doing so, the electrical repulsion between the electrons becomes so great that the objects cannot penetrate further, but are repelled. So solid bodies are solid only because there is electrical repulsion between their negatively charged "electron skins" when they get too close. All collisions, friction, in fact all forces we are familiar with in everyday life, except gravity, are due to electromagnetic forces – even chemical reactions, but that's another story. So ultimately, virtually every motor is an electric motor in the broader sense, unless it is powered by a nuclear reactor, in which case nuclear forces are involved.

Of the basic forces, Newton knew only gravitation at that time. He already knew something about electricity and magnetism, but not enough to be able to formulate a precise law of force. It was to take another 150 years until then. Nothing was known about atomic nuclei. But the idea that the whole variety of natural phenomena arose from simple mathematical laws relating to forces, as Newton defined the term, already existed at that time.

Another insight from Newton's work is that the forces involved in a process always add up to zero. In particular, if only two objects A and B are involved in a process, then they experience opposite forces *(actio* equals *reactio)*. We have already given the recoil of firing a firearm as an example of this principle in Chap. 6. The recoil on the shooter is just as great as the force that hurls the bullet forward, only in the opposite direction. But since the shooter is much heavier, at least he experiences much less acceleration. If he leans on a wall while firing, the recoil will go to the entire earth. But the mass of the earth is so great that the acceleration it experiences due to the force is negligible.

The same is true for gravity: The moon attracts the earth just as strongly as the earth attracts the moon. This causes the Earth to describe small "swerves" in its orbit around the Sun. Since it is about 80 times heavier than the Moon, these swerves are just 80 times smaller than the elliptical orbit with which the Moon moves around the Earth. We don't notice the swings on Earth. But the moon's gravity also makes itself felt through the tides. The side of the Earth that is closest to the Moon at any given time is pulled slightly more strongly by it than the opposite side. As a result, the earth is stretched a little bit into the direction of the moon. The solid part of the earth cannot react to this. The rotation of the earth is too fast for that, it tilts a different side towards the moon every few hours, but a deformation of the rock would take much longer. But the liquid part of the earth's surface, that is, the sea, can react; the water rises in the direction of the moon, and on the opposite side it rises too (away from the moon), and thereby stretches the earth by a few meters.

The sun also swerves a bit due to the gravitation that the planets exert on it. Here, Jupiter imposes the strongest effect, as it is by far the heaviest planet, 300 times heavier than the Earth, but still 1000 times lighter than the Sun.

The **law of conservation of momentum** follows from the rule that the forces in a physical system as a whole cancel each other out (i.e. add up to zero). The **momentum of** an object is the product of its mass and its velocity. The force acting on the object is the derivative (instantaneous change) of its momentum. When the forces of a system add up to zero, it means that the change in the total momentum (i.e., the sum of the momentums of the individual objects) vanishes. Thus, the total momentum is constant ("conserved"). Conservation laws are very practical. On the one hand, they are principles that can be used as a guide. On the other hand, they can often be used in arithmetic to find a "shortcut" to the solution of a problem.

The idea that the world is one big machine, a kind of clockwork, was expressed by some philosophers even before Newton. With Newton's physics, however, the idea was first elaborated on a large scale and given a precise mathematical foundation. A "god" would have to specify only two things: first, the initial conditions at an initial point in time, i.e. the positions, velocities and masses of all particles of matter at that one point in time (with electromagnetism, one would still need the electric charges, but that was not yet known in Newton's time), and second, the general laws of force, such as the law of gravity. Then the course of the entire history of the world could be calculated from this. The movements of all particles at all times are then already "encoded" in a single point in time, in the initial conditions. Just as it is known to be the case with the solar system, the planets rotate invariably as long as they exist, in their motions around the sun, there can be no surprises there (unless a foreign body enters the solar system), everything is fixed for eternity. The surprises we experience are only due to our ignorance of the exact positions of all particles and our lack of computing capacity – as in the example of the die already mentioned, whose throwing result we cannot predict for these practical reasons (Chap. 1).

Energy

The best known of all conservation laws is probably the law of conservation of energy. But it is not so easy to explain what energy actually is. The concept of energy is much more difficult than that of force or momentum. Nevertheless, the term is on everyone's lips because of its practical significance. Energy can be "stored" and "transformed". We say energy is "generated" or "consumed", but these are just colloquial expressions for certain types of energy conversion. We speak of "sources of energy" when we are talking about being able to transform certain forms of energy into other forms that are useful to us.

It takes some mathematics to understand exactly what the term means in physics. Roughly speaking, it makes use of a certain property of a certain type of forces. These so-called conservative forces ("conservative" precisely because they conserve energy) include, in particular, gravity and the electric force. Because conservation laws are so practical, it makes sense to define a quantity mathematically in such a way that it is conserved, even if it is quite obscure, in order to benefit from it later when doing calculations. Energy comes about in exactly the same way. For the conservative forces, one can define two quantities, the "potential" and the "kinetic energy", so that an increase of one always leads to a corresponding decrease of the other, thus their sum is constant (conserved). The kinetic energy results from the velocities of the objects involved, whereas the potential energy results from their positions relative to each other.

The magnetic force is not conservative, i.e., one cannot define a potential energy for it. Fortunately, however, it is of such a nature that it does not change the kinetic energy: It accelerates charged objects orthogonally to their direction of motion, so that only the direction of the velocity changes, not its magnitude. So here, too, the energy remains the same.

Other forms of energy can be traced back to the kinetic and potential ones. For example, "electrical energy"is a shortened term for "potential energy associated with the electrical force". Heat is nothing more than the disordered kinetic energy of many particles (when it's warm, it means the particles are scurrying fast; when it's cold, they're scurrying slowly). Chemical energy is the electrical energy of molecules. In a chemical reaction, molecules are changed into other molecules. According to the law of conservation of energy, if their potential energy is less, it means energy has been "released", i.e. converted into heat (the new molecules scurry faster than the old ones), or, in an explosion, into outward motion. Nuclear energy is potential energy in the nucleus of an atom, associated with both nuclear forces and the electric force between protons in the nucleus). Radiant energy is also kinetic energy, namely that of the particles that make up the radiation. However, because of the "wave-particle duality"of radiation, which we will discuss later, the situation here is a bit more complicated.

The law of conservation of energy is often presented as a deep principle of physics. In classical mechanics, however, it appears more like a mathematical trick that can be applied to a large number of physical laws. The deeper meaning of the concept of energy becomes apparent only in SR and GR as well as in QM. At the same time, however, in GR the law of conservation of energy loses its validity. For example, radiation in an expanding universe loses energy without being converted into another form.

Degrees of Freedom

In the case of extended bodies such as planets, billiard balls or flying knives, in addition to the speed of the body as a whole, one must also take into account the rotation (the earth not only rotates around the sun, but also around "its own axis"). Classical mechanics deals with this as well. Here a bunch of new terms are added, such as angular velocity, angular momentum, torque, and rotational energy (which is also a form of kinetic energy), but they have a lot of similarity to the previous terms. When the body is not spherical, such as in the case of a flying knife, new complications arise depending on which axis the body is rotating about. Interesting gyroscopic movements are the result.

In order to keep the task clear, it is important that the bodies are rigid, i.e. there are no individual parts that flap back and forth as in a human body. Of course, this is a rather rough approximation even for a planet. (In fact, the wobbling back and forth of the tides on Earth, for example, provides friction that slows the rotation of the Earth in the long run). It applies much better to billiard balls.

A **degree of freedom** is a time-dependent variable that describes the state or position of a physical system. A point-like particle has three degrees of freedom, namely its position in three-dimensional space, which is described by three coordinates (usually x, y, z). In the case of a rigid body, there are three further degrees of freedom, which describe its orientation in space.

The number of degrees of freedom determines the number of independent ways a physical system can move or change. Thus, it also determines the number of equations of motion that you must solve to predict the behavior of the system.

We refer to **classical mechanics** as the area of physics in which 1) the definitions and laws established by Newton apply, and 2) the number of degrees of freedom is limited. To keep the number of degrees of freedom manageable, one often has to make certain idealizations (e.g., neglecting the "wobbling" of the planets). How useful these idealizations are depends on the individual case and the concrete problem.

This allows us to distinguish classical mechanics from other areas of physics. If the number of degrees of freedom becomes infinite, we are dealing with field theory. If the number remains finite, but is so huge that we are only interested in certain statistical quantities, we are dealing with statistical mechanics. In QM, Newton's rules lose their validity on microscopic scales. In the two theories of relativity, space and time are combined in a new way, so that numerous Newtonian definitions and rules have to be modified.

7.2 Classical Field Theory

Classical mechanics deals with systems with finitely many degrees of freedom: a small number of "point-like" imaginary particles with certain masses (fixed once and for all) moving through space and described by their positions and velocities, or else solid extended bodies for which additionally their shape (fixed once and for all) and their orientation in space have to be specified.

The situation is different for fluids (that is, liquids and gases). Although these ultimately also consist of small particles (atoms or molecules), firstly it is not practical to calculate with the individual positions and velocities of such a huge number of particles, secondly hardly anything was known about their properties until about 1900, and thirdly Newtonian mechanics is not even applicable to many aspects of these particles, because the laws of QM apply here. Already in the eighteenth century, fluids were studied in detail and described by physical equations without having to refer to any particles.

Fluids fill certain areas of space. In contrast to solids, they can be deformed at will and are partially compressible (liquids only to a small extent, gases much more so). To account for this, one must extend the concepts of mechanics somewhat. One imagines the volume filled by the fluid divided into small units. Instead of the total mass, one considers the **density,** or mass per unit volume. This can vary from place to place (only slightly for liquids, as I said, and more for gases) and can change with time. It is therefore a function of position and time, whereas in mechanics everything was a function of time only. Density is denoted by the Greek letter ρ, so as a function it is written $\rho(x, y, z, t)$ because it depends on the three coordinates of location and time. For example, we know that the Earth's atmosphere gets thinner and thinner as we go up. So if z is the coordinate that represents the height above the ground, then ρ continues to decrease as z increases.

When fluids move, they are said to *flow.* The velocity of flow can also depend on the exact position and vary with time. It is denoted by $\vec{v}$, and as a function it reads $\vec{v}(x,y,z,t)$. The small arrow means it is a **vector;** it has a magnitude and a direction, unlike the **scalar** quantity ρ, which has only a magnitude. In ρ and $\vec{v}$, we are dealing with **fields;** this term is used for all quantities that depend on location and time. A field has infinitely many degrees of freedom: its value can vary separately at any point in space – not completely independently, the physical equations take care of that, but still in such a way that infinitely many numerical values are needed to describe the exact situation at a given time.

Density and flow behavior are influenced by different types of forces described in the theory of **hydrodynamics.** A river flows from its source to the sea, ultimately driven by gravity acting on each unit volume of water, but also by the **pressure** (a very important concept in hydrodynamics) of the downstream liquid pushing the whole thing forward. **Viscosity** also plays a role, a measure of how strongly adjacent layers of fluid apply friction on each other. The slower layers thus try to stop the faster ones, the faster ones try to drag the slower ones along. This can be observed particularly well with honey, a liquid with a very high viscosity.

The equations that describe this behavior in general form are called the **Navier-Stokes equations.** They are decidedly complicated; they are *coupled partial differential equations:* Differential equations because they contain derivatives; coupled because the equations depend on each other, you can only solve them all together, not individually; partial because, unlike in mechanics, the derivatives describe not only temporal but also spatial dependencies. Changes occur in different directions: Density and flow velocity change with time, but also vary with location. In a river, water flows faster in the middle than near the banks, faster at the surface than near the bottom. For each direction, one must form a separate derivative, i.e. calculate a separate change, each of which describes only part of the overall behavior, hence the term "partial".

The equations are so complicated that they are relatively difficult to solve in certain situations, even with approximation methods, and that is when *turbulence* occurs, small vortices or eddies, as you can observe in flowing waters near rapids; in the air, you know this from the turbulence you experience in an airplane. These eddies and vortices are very difficult to predict accurately. From the mathematicians' point of view, it is not even clear whether the equations in such cases have solutions at all. To clarify this question is the content of one of the seven Millennium Problems, for which a prize of one million dollars is offered. We have already discussed this in Sect. 6.3 and also made clear why a physicist need not be particularly concerned about it.

Waves

A typical phenomenon in field theory is the propagation of **waves.** We are most familiar with water waves, for example in the sea: Due to the various forces acting there, the water rises and falls. *Wave crests* (places where the water is highest) and wave *troughs* (places where the water is lowest) are formed. The wave crests move in a certain direction. It is not the individual water particles that move with the wave. The individual water particles essentially only move up and down, not forward. The movement of the crests of the waves is caused by the water rising at some points and falling at others at any given moment (the sinking of the water at one point pushes it upwards at the neighbouring point), in such a way that the point where it is currently highest moves relatively evenly in a particular direction. Two important terms are wavelength and frequency: **wavelength** is the distance between two wave crests, **frequency** is the number of wave crests that drift past a point within a certain time interval.

In the case of **sound waves,** it is the pressure which alternately assumes higher and lower values. Overpressure and underpressure alternate at short intervals, and are passed on by the forces that overpressure and underpressure exert on their immediate surroundings, with a certain speed, the *speed of sound,* which depends on the medium, i.e. the substance, in which the sound propagates. In air, it is slightly more than 300 meters per second.

Sound waves are typically triggered by vibrations acting on the medium (e.g. air). This is done particularly artistically with musical instruments. These are constructed in such a way that they each vibrate in a very specific characteristic way. The vibration acts on the air, causing rapid sequences of overpressure and underpressure at the edge of the instrument, which propagate through the air as sound waves, where they can then cause another piece of solid matter to vibrate elsewhere, for example our eardrum, causing us to *hear* the music. It is similar with speech: The vibration of our vocal cords combined with the shape we give our lips and tongue at that moment creates a specific pattern of sound waves that leave our mouth, propagate through the air, and vibrate something again elsewhere, such as an eardrum or a microphone.

This also shows that waves are suitable for transmitting information. A *transmitter* transmits its own vibrations, whose specific pattern encodes an information, to the medium, which carries this information in the form of a wave to one or more *receivers,* where it is converted back into vibrations from which the information can be read. In our brain, the vibration of the eardrum is translated into the sensory perception of sounds and the recognition of speech. The frequency of the sound waves corresponds to the perceived pitch. The "concert pitch" A, for example, corresponds to a frequency of 440 hertz, i.e. 440 overpressure phases per second. Each octave doubles or halves the frequency.

Another important property of waves is that they can be **superimposed;** the technical term for this is **superposition** or **interference.** This means that waves emanating from different sources do not interact with each other. They superpose in the sense that the respective positive and negative pressures of the individual waves simply add up. At the receiver, e.g. our eardrum, there is now a vibration pattern that just corresponds to the sum of the individual waves. If an overpressure arrives from each of two waves – let's assume the same strength in both cases – they add up to twice the total overpressure, the eardrum is deflected twice as much from its resting position as it would have been with each wave separately *(constructive interference).* If, on the other hand, an overpressure arrives from one wave and a negative pressure from the other, they cancel each other out and the eardrum is not deflected at all *(destructive interference)*.

It is crucial for the transmission of information that the individual components can be filtered out again at the receiver, although there is only one vibration pattern there that corresponds to the sum of the individual patterns. Our brain is amazingly good at this. Although each individual musical instrument already produces a superposition of vibrations of different frequencies – that is precisely what makes up the specific sound of each instrument – we have the ability to listen to an entire orchestra and thereby pick out the individual instruments from the oscillations of just two eardrums, even just one if we are deaf in one ear. From tiny differences in the pattern of vibration we are able to clearly identify the voice of an acquaintance, we can even filter it out in a crowded room from the simultaneous murmurings of several people, again from the oscillation of only two eardrums.

Force Fields

From school lessons we know the term *field* in physics mainly as a *force field*. In Newton's theory of gravity, each body acts on every other body according to its mass, exerting a gravitational force. Many found it difficult to imagine this force acting across empty space from a distance, so each body was said to permeate all of space with a *force field (gravitational field)* in which the other bodies then move. In this conception, it is no longer the distant other body B that exerts the force on a given body A, but the force field at the point where A is currently located. This, however, is a mere idea formed by our imagination. For the equations describing the motions of bodies with the help of Newtonian gravitation, it makes no difference whether we imagine it as acting directly from a distance or transmitted through a force field.

This changes in GR, the gravitational theory of Einstein, which is more precise than that of Newton. GR is clearly a field theory, a direct effect over the distance does not exist in it. Changes in one place (for example, a motion of body B) only propagate gradually through space by means of a field, at the speed of light, and it is the field at the position of A that exerts an effect on A. What this field is, we will discuss in Sect. 7.4.

One of the most important field theories is the theory of **electromagnetism** by James Clerk Maxwell. The starting point of this theory is that there are **electric charges**, an abstract property of matter, a numerical value that can be assigned to any piece of matter. These charges, by their presence at a certain place and additionally by their movements, influence two vector fields, i.e.

two physical quantities, each of which has magnitude and direction and is mathematically represented by functions of position and time. These vector fields are called **electric field** and **magnetic field.** One can characterize these fields as abstract properties of space: Every point in space at every instant has two properties, each of which can be expressed by a magnitude and a direction, and these two properties are called the electric field and the magnetic field. These fields in turn exert forces on all electric charges. So there is a mutual influence: the charges influence the fields, and the fields in turn influence the charges. Via the fields, therefore, the charges can mutually influence each other: Two positive charges repel each other, as do two negative charges. A positive and a negative charge, on the other hand, attract each other. The fields also influence each other: a change in the magnetic field over time has an effect on the electric field and vice versa.

The exact behavior of these interactions is described by **Maxwell's equations.** As in hydrodynamics, these are coupled partial differential equations, but fortunately somewhat simpler than the Navier-Stokes equations. There is a large variety of solutions to these equations. Innumerable electrical and magnetic phenomena can be derived from them, innumerable technical applications can be constructed from them.

The solutions of the equations also include the **electromagnetic waves** that propagate through space at the speed of light, about 300,000 kilometers per second, about a million times faster than the speed of sound in air. In the case of these waves, it is the magnitudes of the two fields that are subject to perpetual ups and downs. They include what we perceive as light. More precisely, it is a small subrange of the **electromagnetic spectrum,** that is, of the entire range of possible wavelengths of electromagnetic waves that we perceive as **light,** namely the range between about 400 and 800 nanometers. The exact wavelength is what we recognize as *color,* 400 nanometers we see as violet, 800 nanometers as red. This is why the invisible region above 800 nanometers is called *infrared,* and the invisible region below 400 nanometers is called *ultraviolet.* Even the invisible regions of the spectrum have certain effects on us or can be used for certain technical applications. Therefore, certain regions of the spectrum have names that we associate with these effects and applications: X-rays, microwaves, radio waves, and so on.

The same applies to electromagnetic waves as to sound waves: They can be used to transmit information. It works even better with them, because electromagnetic waves often have a very long range, unlike sound. They are also transmitted through airless space, which is why we can see the stars but not hear them. If the wavelength is long enough (radio waves), they also penetrate walls and other obstacles. A transmitter makes charges vibrate, which leads to the emission of the waves. A receiver converts the waves back into vibrating charges and filters out the portion it is interested in, such as a particular radio station or telephone conversation. For the principle of superposition also applies to these waves: all radio frequencies, all telephone calls of the mobile phone networks that are accessible in our environment flow through us at the same time; our devices only have to know how to get to the portion intended for them.

7.3   Special Theory of Relativity

SR is by far the simplest theory in physics. It can be written down in a few lines. Middle school math is enough to understand it. You do not need to solve differential equations, it is enough to form squares and draw roots.

So why was it discovered so late, in 1905? There are two reasons for this. First, its effects only become visible when we are dealing with velocities close to the speed of light. Secondly, it contradicts our intuition. Already in school physics we learn that one must not compare "apples with oranges", i.e. that e.g. spatial distances, time and mass are three different physical quantities which must not be set off against each other and are provided with different units (meter, second and kilogram in civilized countries; an archaic muddle of traditional units of measurement in the Anglo-Saxon ones). SR now says, contrary to these warnings, that space and time are to be combined with each other after all. Probably it really took a genius outsider like Einstein to dare to do this.

SR became necessary after Michelson and Morley obtained a strange result in a sophisticated experiment in 1887. It is well known from road traffic that in a head-on collision the velocities add up, but in a rear-end collision they subtract, which is why the former has much worse consequences than the latter. In a head-on collision, if one car is going 100 km/h and the other is going 70 km/h, then they collide at a speed of 170 km/h relative to each other. In a rear-end collision, on the other hand, if the rear car is going 100 km/h and the front car is going 70 km/h, then the relative impact speed is only 30 km/h. Roughly speaking, Michelson and Morley found that this rule does not apply in the case of the speed of light. If you are running away, with velocity v, from a beam of light moving towards you at the speed of light c, then the beam of light should hit you at a relative speed $c - v$, as in a rear-end collision. On the other hand, if you are running towards it with velocity v, the relative velocity should be $c + v$ as in a head-on collision. However, in the experiment, it came out that the relative velocity is c again in both cases. The simplest rules of addition and subtraction no longer seemed to apply!

18 years later Einstein delivered the explanation with SR. In it, he elevated the constancy of the speed of light to a principle and, starting from this, made all kinds of considerations, with mirrors, moving trains, signals flying back and forth; with people comparing clocks and trying to agree on what "simultaneously" means. With this, he finally came up with the structure of spacetime, which all of this came down to. Some of these considerations are quite complicated, but they still dominate popular scientific accounts of the subject.

I would like to go a different way here. Once the structure of spacetime described by SR is understood, I think it is easier to take it as a starting point in describing the theory, rather than the constancy of the speed of light. The constancy of the speed of light, as well as all other aspects of the theory, then follow from this structure. Surprisingly, this simpler route is not taken anywhere (that I am aware of) in the literature. The following presentation therefore possibly celebrates its premiere here.

Perhaps it is also a matter of taste which way is easier. The structure of spacetime is quite a mathematical matter, whereas the constancy of the speed of light is what we call a **physical principle.** It may therefore be that mathematicians generally tend to find the former easier and physicists the latter. But it is also a question of which of the two is to be regarded as more fundamental.

In my eyes, the preferred view is that the constancy of the speed of light follows from the structure of spacetime and not vice versa. The structure of spacetime is more fundamental. In general, there seem to be two possitions among theoretical physicists: Some find physical principles very important and fundamental (the conservation of energy, the constancy of the speed of light, etc.) and want to use them as a starting point in any theory as far as possible. The others think that such principles are rather consequences of the mathematical structures of a theory and that it is misleading to put them as quasi "God-given" at the beginning. Einstein belongs to the first group, I to the second.

The American philosopher of science Tim Maudlin elaborates on the second point of view in an interview. He refers to the "mother of all principles", the conservation of energy, as an example:

It is not that the principle is literally a fundamental axiom of any physical theory. [...] The confidence that one has in the principle arises from the fact that it works across a very wide field of application. Sometimes it can even be derived. [...] But ultimately, the fact that the maxim works is an explanandum, not an explanans: *it is something to be accounted for by the fundamental dynamics.* (Tim Maudlin, quoted in Schlosshauer, 2011, p. 209)

This is similar to what I already said in Sect. 7.1 about the conservation of energy. But back to SR.

The Two Postulates of the SR

The structure of **spacetime** can be summarized in two postulates:

1. Space and time form a common four-dimensional continuum (three dimensions for space, one for time). They can therefore be expressed in the same units. That is, you can convert seconds to kilometers in much the same way that inches are converted to centimeters. The conversion factor is: 1 s = 299,792 km.
2. There is only one difference between the time and space dimensions: If, in a right triangle, one cathetus is in the direction of time, then the Pythagorean theorem applies with minus signs instead of plus signs (Fig. 7.1). In this way distances of points in spacetime are defined.

At this point, for once and the only time in this book, I would like to demonstrate some small calculations to illustrate these two points. You will see that the calculations are very simple, otherwise I would not dare to do this either, but would be afraid to lose my readers. The two statements about the structure of spacetime strain our intuition (converting seconds into kilometers! Wrong Pythagorean theorem!), but not our calculation skills.

From the first postulate it follows first that velocities are "dimensionless" quantities, i.e. numbers without units of measurement. Normally, velocities are given in meters per second (or in units derived from them, such as

(i) $c^2 = a^2 + b^2$ (ii) $c^2 = b^2 - a^2$ (iii) $c^2 = a^2 - b^2$

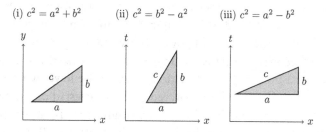

Fig. 7.1 Modified Pythagorean theorem in spacetime. (i) "Normal" Pythagorean theorem in two spatial dimensions. (ii) and (iii) Modified theorem with a cathetus *a* in space and a cathetus *b* in the time direction, where *b* is longer than *a* in one case and shorter in the other

kilometers per hour). You can convert the second to meters according to the first postulate. You then get "meters per meter", which is to be understood as "meters in the direction of space per meter in the direction of time". This is analogous to the gradient of roads. If you see a road sign with a warning about a 10% slope, it means that when you go ten meters in the horizontal direction, it goes up one meter in elevation at the same time. There, too, you have the unit "metres per metre", this time to be understood as "vertical metres per horizontal metre". The unit of meters cancels out; what matters is that it goes up 1/10 as much as it goes forward, which means the same as 10%. Exactly the same now applies to velocities. A speed of 1/10 means that for every ten meters in the future direction, it goes forward one meter. You can do the same reasoning in units of seconds. Similarly, you can say that a speed of 1/10 means that for every ten seconds in the direction of the future, it moves forward one second in space, or 299,792 kilometers. The confusion you may feel at such sentences comes only from the fact that we are not used to using meters and seconds as equivalent units, differing only by a conversion factor, like inches and centimeters. But that's just the point of SRT: space and time are equivalent, except for the Pythagorean theorem thing.

To get a feel for these numbers in realistic cases: What numerical value does the speed of 30 meters per second correspond to? Well, 30 meters per second means as much as 30 meters divided by 1 second, means as much as 30 meters divided by 300,000 kilometers (we round a tiny bit to simplify the calculation), means as much as 30 meters divided by 300,000,000 meters, means as much as 1/10,000,000. So the speed 30 meters per second is the same as the speed one ten millionth (without a unit of measurement).

The speed with the value 1 has a special name. It is called the **speed of light.** This is because light moves with this speed. In my account, the knowledge that light moves at this speed is at the end. Therefore, the use of the name speed of light for the speed with the value 1 is an anticipation of future knowledge. An object with this velocity (i.e. a ray of light, for example) travels a distance of one second, or 299,792 kilometers, in a time interval of one second. Above I called the speed of light c. So my statement is that $c = 1$.

Now to the second postulate, the matter of the Pythagorean theorem. Let us first consider a purely spatial situation. Suppose we have two points A and B in a plane. The plane has been covered with a Cartesian coordinate system ("*x-axis* and *y-axis*"), such that A is at the origin, that is, at the point with coordinates (0,0). Suppose that in this coordinate system, B has coordinates (3,4). That is, to get from A to B, we must go three units of length in the *x-direction* and four units of length in the *y-direction*. According to the Pythagorean theorem, the distance s of points A and B is given by $s = \sqrt{3^2 + 4^2} = \sqrt{25} = 5$.

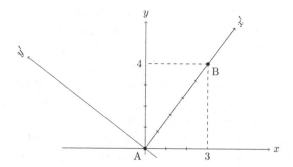

Fig. 7.2 If the coordinate system is rotated, distances are preserved and can be calculated according to the same rule. Here B has coordinates (3, 4) in the coordinate system (x, y), and coordinates (5,0) in the coordinate system (x´,y´). The distance to A is $5 = \sqrt{3^2 + 4^2} = \sqrt{5^2 + 0^2}$

The distance s is of course independent of what kind of coordinate system we have chosen. We can rotate or move the coordinate system. Then the coordinates of points A and B change, but their distance remains the same. For example, we can rotate the coordinate system so that B is on the *x-axis*. Then B has the coordinates (5,0), and s *is* calculated by the same method to $s = \sqrt{5^2 + 0^2} = \sqrt{25} = 5$ (Fig. 7.2).

This can be made a little more precise and generalized. A rule for calculating the distance from the coordinates of two points is called a **metric.** Exactly what the metric looks like generally depends on two things: the geometric properties of the space and the coordinate system[1] chosen . In a Euclidean space – that is, a space in which the Euclidean axioms of geometry apply – and with a Cartesian coordinate system – that is, a coordinate system with straight axes perpendicular to each other – the following applies: If A has coordinates (x_1, y_1) and B has coordinates (x_2, y_2), then the distance s from A and B is

$$s = \sqrt{\left(x_2 - x_1\right)^2 + \left(y_2 - y_1\right)^2} \tag{7.1}$$

The distance s is independent of the choice of the coordinate system. As a physicist, one says: It is **physical,** which should mean as much as: It does not depend on my own chosen conventions. The coordinates themselves, of course, depend on the choice of coordinate system; they are not physical. Equation (7.1), which is used to determine s from the coordinates, depends in

---

[1] Note for mathematicians: Yes, you can also define a metric coordinate-independently. But here I am explicitly talking about the coordinate-dependent version.

part on the coordinate system: namely, it applies only to *Cartesian* coordinate systems. Thus, if you change from one Cartesian coordinate system to another Cartesian coordinate system, the coordinates change, but the equation that links *s* and the coordinates remains the same, as in our example. If, on the other hand, you change from a Cartesian coordinate system to a polar coordinate system (you know this from the picture of a radar screen: each point is characterized by its distance from the center and by a direction, i.e. an angle), both the coordinates and the relationship between *s* and the coordinates change. Two Cartesian coordinate systems are always related by rotation, translation and/or reflection. It follows that rotations, translations and reflections are exactly the operations that can be applied to a Cartesian coordinate system without changing the metric.

Now we replace the *y-axis* by a time axis *t*, i.e. we consider a plane with a space dimension *x* and a time dimension *t* (forgetting for simplicity the other two space dimensions for the moment). This time we assume that A has coordinates (0,0) and B has coordinates (4,5). A and B are not points in space, but points in spacetime. The first coordinate describes the spatial position along a straight line, the *x-axis*, and the second coordinate defines the point in time. In SR jargon, such points in spacetime are called **events.** The two coordinates define the where and when of the events: event B takes place at the position with coordinate 4 and at the time with coordinate 5. This means, by the way, that you have to move with 4/5, i.e. 80% of the speed of light, to get from A to B (five units in the future, four units in the *x-direction*, which makes a speed of 4/5). Whether the units here are meters or seconds or inches doesn't matter, essential is only that you take the same unit for the spatial distance and the temporal distance. The second postulate now states that a spacetime distance *s* is defined for A and B, using a modified Pythagorean theorem: $s = \sqrt{5^2 - 4^2} = \sqrt{9} = 3$ (Fig. 7.3). So the total distance is *smaller* than the two coordinate distances because of the minus sign. In space alone this would be unthinkable!

This kind of distance is thus doubly unusual: first, we are used to considering spatial and temporal distances separately. However, the distance defined here refers to a kind of "diagonal" across space and time. Secondly, the total distance is smaller than the two perpendicular partial distances of which it is composed, which to our mind sounds "impossible" or ill-defined. Yet this distance is the crucial measure in SR and the reason, the logical cause for all its effects. Again, it is true that the distance *s* is independent of which coordinate system we choose. (This is what is meant in the postulate, it is supposed to be a *physical* distance).

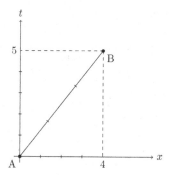

Fig. 7.3 By the modified Pythagorean theorem, the distance between events A and B is not $\sqrt{5^2+4^2} \approx 6,4$, but $\sqrt{5^2-4^2} = 3$

Time Dilation

One can regard a coordinate system as the **reference system of** an observer. Here we assume that the observer thinks "egocentrically" and places himself in the center of a coordinate system. The coordinate system refers to him, so it is called his frame of reference. "Placing himself in the center" means that he claims for himself the spatial coordinate $x = 0$ at all times. The time axis is the set of all points with $x = 0$; it consists of all events that take place at the observer's location, so in a sense it is the observer's "here". (Put a little more flippantly: The time axis is always where the observer is).

Let us assume that two observers, Erwin and Otto, are moving relative to each other with constant velocity. From Erwin's point of view (in Erwin's frame of reference), he himself is at rest. Otto approaches him, meets him at a certain point in time, and then moves away from him again without stopping. From Otto's point of view, Otto is at rest, Erwin approaches him, encounters him, and moves away again. Let's further assume that the coordinate system from our example above is Erwin's frame of reference. The two meet there at point A. After the encounter, Otto moves from A to B, relative to Erwin thus with 80% of the speed of light.

The time axes of the two frames of reference are tilted against each other: they approach each other, intersect in the event of the encounter (the only moment when the here of the two is identical), and move away again from there, just like the observers themselves. In Otto's frame of reference, B lies on his time axis – we call it t', to distinguish it from Erwin's time axis – because it is, after all, an event in Otto's "here" (Fig. 7.4).

The coordinates of B are in Otto's reference system $(0, \tau)$, where the value τ is still to be determined. It is determined by the fact that with the "modified

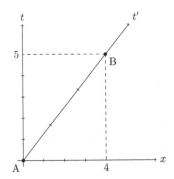

Fig. 7.4 Otto's time axis runs through A and B. Due to the modified Pythagorean theorem, it appears to be stretched out to a certain extent ("time dilation")

Pythagoras" we must have s = 3 again, because s is independent of the reference system, $s = \sqrt{\tau^2 - 0^2} = \sqrt{\tau^2} = \tau$. So τ = 3. And more generally: the **proper time τ of** a uniformly moving observer always corresponds to the spacetime length s, which he has travelled. So Otto has travelled only three time units between A and B, but for Erwin it was five time units, between the same two points! For Otto, less time passes between events A and B than for Erwin! This phenomenon is called **time dilation** and has actually been observed. Here we could derive it directly from the postulates about the structure of spacetime, without using the constancy of the speed of light and talking about signals running back and forth (as all other popular science books I know do).

This representation also has the advantage of being able to emphasize a certain aspect: In the "conventional" representations, we are told that space and time distances are relative, i.e. depend on the frame of reference. Here I have explained that there is something which is not relative but absolute ("physical", independent of the frame of reference), namely the spacetime distance s. And just the absoluteness of s is the reason, why space- and time-distances must be relative each by itself. In my opinion, this aspect is very important.

If Otto turns around at point B and returns to Erwin with 80% of the speed of light, he will meet him again at point C, where C has the coordinates (0,10) in Erwin's frame of reference. The spacetime distance of B and C is again 3, which is again Otto's proper time on this path. So Erwin has aged ten time units between A and C, Otto only six (Fig. 7.5). Fast travel keeps you young. This phenomenon is known as the **twin paradox.** (It was originally told as a story of two twins who, after such a journey, were suddenly no longer the same age).

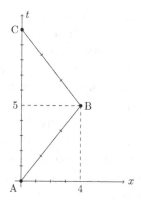

Fig. 7.5 "Twin paradox". Otto has aged by six time units on the way A-B-C, Erwin on the direct way from A to C by ten. In the direction of time, the direct path is not the shortest but the longest

Lorentz Transformation

The general metric for points with coordinates (x_1, t_1) and (x_2, t_2) is:

$$s = \sqrt{\left|(x_2 - x_1)^2 - (t_2 - t_1)^2\right|} \tag{7.2}$$

The absolute value bars here mean that the expression under the root is to be reversed, i.e. $(t_2 - t_1)^2 - (x_2 - x_1)^2$ if the time difference $|t_2 - t_1|$ is greater than the spatial position difference $|x_2 - x_1|$, as in our example. After all, you can't take the root from negative numbers. Again, we can now consider what coordinate transformations we can apply without changing the form of the metric (Eq. 7.2). In the Euclidean metric, these were coordinate transformations that came about by translation, mirroring, or rotation. Here, however, we are dealing with a space and a time axis, and one can calculate that conventional rotations of such a coordinate system change the metric. Instead, so-called **Lorentz transformations** work, which I will characterize in a moment.

The diagonal (bisector) between the *t-axis* and the *x-axis* of Erwin's reference frame consists of the points with the same coordinates, $x = t$. Any two points on this diagonal have spacetime distance zero, because it is $(t_2 - t_1) = (x_2 - x_1)$, and thus the expression under the root disappears. This is also unusual. In "normal" metrics, such as Euclidean, the statement "A and B have zero distance" is identical to "A and B are the same point". In the strange metric of SRT, this is no longer true. The path between two points on the diagonal travels at the speed of light: for every unit in the *x-direction*, it also travels one unit in the *t-direction*. This means that no matter how fast light travels, it only ever covers the spacetime distance zero. This can also be expressed in the way that time dilation causes the travel time of light (in the reference frame of light) to shrink to zero. That is, from our point of view, light takes about 8 min to travel from the sun to the earth, covering a spatial distance of 150 million kilometers. If we could ride on a beam of light, however, we would find that from this perspective it takes not 8 min, but exactly zero seconds, and in that time we cover a distance of length zero. This is also the reason why a higher speed than the speed of light is purely logically impossible as long as SR is valid. The speed of light already makes everything shrink down to zero. More is not possible.

The spacetime distance s of two points is independent of the reference frame. Since spacetime distance zero always means speed of light, the principle of constancy of speed of light follows from this, which was the starting point for Einstein (both from his theoretical considerations and from the experimental situation, after the Michelson-Morley experiment): Spacetime distance zero remains spacetime distance zero, in every reference frame.

For the Lorentz transformation we do not only demand that s is zero also after the transformation, but s must still result as zero from Eq. 7.2. One can (with a little knowledge of mathematics) quickly conclude that for this purpose the diagonal between the "old" axes must still be the bisector between the "new" axes in the new coordinate system. With the transformation into the reference system of Otto, we have seen that the t-axis is tilted in comparison to the original t-axis. For this to become a Lorentz transformation, the x-axis must be tilted in the opposite direction, toward the t-axis, in order to get the bisector as desired (Fig. 7.6). A slightly more detailed calculation (which I will

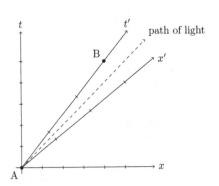

Fig. 7.6 Lorentz transformation. (x, t) is the reference frame of Erwin, (x', t') that of Otto on his way from A to B. A ray of light travels along the diagonal (or angle bisector), in both reference frames

spare you here) shows that all other spacetime distances then follow correctly again using the metric (Eq. 7.2). So the Lorentz transformation tells Otto how to set his x'-axis (the t'-axis is given by the motion), so that the correct distances follow from Eq. 7.2 for him, too.

The x'-axis is the set of all points with $t' = 0$, i.e., from Otto's point of view, the events on the x'-axis take place simultaneously (however, only in a very abstract sense, because in reality signals, for example light impulses, which are emitted from somewhere on the x'-axis, take a certain time to reach Otto, i.e., he does *not* experience them simultaneously; only if he calculates back the running times of the signals, he comes to the conclusion that they originate from the point in time $t' = 0$). Since the x'-axis is tilted compared to the x-axis, this means Otto and Erwin don't have the same idea of what "simultaneous" is. **Simultaneity is relative, depends on the frame of reference.** This contradicts our normal conception of time, in which there is a definite present, a clear demarcation between past and future.

The metric (Eq. 7.2) allows us to distinguish between three types of distances: A distance between two events (x_1, t_1) and (x_2, t_2) is called **space-like** if the spatial distance $|x_2 - x_1|$ is greater than the temporal distance $|t_2 - t_1|$. When the temporal distance is greater, it is called **time-like.** If the spatial distance and the temporal distance are equal, it is called **light-like.** Whether a distance is space-like, time-like or light-like is independent of the reference system.

This distinction enables us, if we wish, to say: we use seconds to measure the length of time-like distances, and meters for space-like distances. This reintroduction of the different assignment of meters and seconds has the advantage of being better adapted to our measuring instruments. Our scales are now calibrated in meters and our clocks in seconds. The disadvantage is that the speed of light is no longer equal to 1, because the equality 1 s = 299.792 km is discarded. This gives all the formulas in SRT numerous factors with powers of c (c, c^2, c^3, c^4), which makes them harder to read and harder to understand, and the *c-factors* ultimately serve only to accomplish the conversion from meters to seconds and vice versa "through the back door", so to speak. Because of these advantages and disadvantages, theoretical physicists usually set $c = 1$ (i.e., affirm that meters and seconds can be converted into each other, thereby avoiding the *c-factors* in the formulas), whereas experimental physicists do not. This is an example of the more common phenomenon that theorists and experimenters use different conventions, so a certain amount of "translation" is needed in a dialog. Theorists prefer short, concise equations that make the mathematical structure clear without being "watered down" by unnecessary factors. Experimenters prefer equations that are adapted to familiar measurement systems, even if that means they have to put up with extra conversion factors. I am a theorist, so it remains $c = 1$.

For space-like distances the following interesting statement holds: If the distance between two events, P and Q, is space-like, then reference frames can be found in which P is temporally prior to Q, as well as reference frames in which Q is temporally prior to P, and those in which P and Q are simultaneous. In other words, the temporal order of P and Q is arbitrary. This means that if it were possible to send a signal from P to Q, then there would be reference frames in which the signal arrived before it was sent off, which is generally considered contradictory. Or more generally, if there were some causal relationship between P and Q, for example in the way that P is to be regarded as the cause of Q, then there would be frames of reference in which the effect occurs before the cause. Therefore, most assume that not only is it impossible to travel from P to Q, but that there can be no signal exchange in general, and no causal connection between P and Q at all. Causal connections always run along time- or light-like distances.

Mass and Energy

So far we have only dealt with distances and velocities within the framework of SR. What about general dynamics, i.e. forces, momentum, masses, energies? In Newtonian mechanics, all these concepts ultimately emerged from velocity and its change, acceleration: a force is proportional to the acceleration it causes, and the mass of the accelerated object is the factor belonging to this proportionality. Mass times velocity, in turn, defines momentum. In SR, the problem with this is that you can't just add velocities together. After all, the Michelson-Morley experiment already shows that $c + v$ is c again. For the same reason, acceleration doesn't work as it used to: if you try to accelerate an object that's already moving at the speed of light by applying a force to it, it doesn't make the object go faster. (Another reason why speeds greater than c *are* not possible!).

The question is whether new definitions can be used to generalize the relations between the concepts of force, momentum, mass, energy on the one hand, and velocity or acceleration on the other, in such a way that other useful relations from Newtonian physics can be carried over into SR, in particular the laws of conservation of energy and momentum. In doing so, consistency must be ensured: The new definitions and relations must be valid in all **inertial frames**, i.e., in all uniformly moving reference frames. This is because all inertial frames are related by Lorentz transformations, and SR regards them all as equivalent: If one passes from one system to another by Lorentz transformation, the form of the equations must be preserved, even if the numerical values change (similar to the discussion of the metric above).

It turns out that this is possible. For this, however, mass must be included in the concept of energy: The **relativistic energy of** an object is defined as Newtonian mass plus kinetic energy. (The formula for kinetic energy is different from Newton's, but it gives almost the same values for small velocities). At the same time, the definition "momentum is mass times velocity" must also be modified to "momentum is relativistic energy times velocity". Therefore relativistic energy is also called **relativistic mass**. This is just the statement of Einstein's famous formula $E = mc^2$, which in our case, because $c = 1$, simply reads $E = m$. Newtonian mass is also called **rest mass** in this context because this is the value of relativistic mass in a frame of reference where the object is at rest and therefore has no kinetic energy. In today's theoretical physics, however, one has mostly returned to using the term mass and the symbol m simply to refer to the rest mass, while no distinction is made between relativistic mass and relativistic energy, because both are the same thing anyway. Both are simply called energy in this context and denoted by the symbol E. Einstein's famous formula thus becomes the triviality $E = E$. This is an example of how great knowledge can be completely absorbed by the terminology that goes with it.

Massless objects, i.e. objects with zero rest mass, are a special case. For them, consistency can only be achieved if one assumes that they move at the speed of light. Here the circle closes: Photons, the elementary particles of light, are massless, therefore they *must* move with the speed of light and thus justify this term. Or vice versa: From the fact that light moves with the speed of light, one can conclude that light is massless.

In this section, apart from the strange postulates of SR, I have also presented you some calculations and mathematical argumentations which deal with squares, roots, absolute values and coordinate systems. I hoped it was acceptable to expose you to this. Without this mathematics, in my opinion, the full consequence of the structure of spacetime cannot really be made clear.

Space and time form a common continuum. Only the strange modified theorem of Pythagoras establishes a difference between space and time and is at the same time the reason for the special properties of the speed of light and the strange relativity of duration and simultaneity. In further logical steps it also enforces the equivalence of mass and energy.

I also discussed some different views among physicists, e.g. the different preferences of theorists and experimentalists, and the different derivation of a theory depending on whether one starts from a physical principle or from a mathematical structure.

One more remark at the end: I have often heard the prejudice that SR describes only inertial frames, but no accelerated frames of reference. For accelerated reference systems then GR is responsible. But this is wrong. It is true that inertial frames of reference are the natural starting point in SR (as in Newtonian mechanics, by the way), because the laws of nature and the structure of spacetime can be represented here in a particularly simple form. But of course, there's nothing stopping you from switching to the coordinates of an accelerated frame of reference. You don't need a new theory for that, it's just a coordinate transformation after all. How much time passes for an accelerated observer, what distances he measures, what the energy balance is for him, etc., all that can be easily calculated with SR. With GR, gravity and the curvature of space-time come into play; but this has nothing to do with accelerations in general, only with the special accelerations caused by gravity.

7.4   General Theory of Relativity

GR was Einstein's second stroke of genius. While the SR follows more or less compellingly from the result of the Michelson-Morley experiment, GR was initially a pure thought construct. Like SR before, Einstein derived it from a *physical principle*. This time the principle was related to gravity. Gravity is the only force that accelerates all objects equally. The fact that a sheet of paper falls to the ground more slowly than a rock is due solely to air resistance, which slows the sheet down much more. In a vacuum, all objects would fall at exactly the same rate. In the case of the electric force, however, the acceleration depends on the ratio of mass and electric charge, and with the other forces it is even more complicated. Einstein's consideration was now: If all objects, as long as there is no resistance, fall exactly the same, then the free fall represents in some sense a natural "normal state", whereas in SR and also already with Newton the uniformly moving reference frames represent the "normal state". There, the deviation from the normal state, i.e. from the uniform motion, must be explained by a force. According to the new theory, the deviation from free fall must be explained by a force instead.

We stand on the ground because the firmness of the ground exerts a force on us that is exactly opposite to gravity and thus keeps us from free falling into the bottomless pit. If I stand on a scale that shows my weight as 80 kg, this means that there is a force of 785 Newton between me and the scale. In fact, the scale measures a force (expressed by default in units of newtons), not a mass, but its display kindly assumes that this force is due to gravity on the surface of the earth, and based on this assumption translates the force into a mass (expressed in kilograms). Now, I can interpret this situation in two ways: 1) it is gravity pushing me *down* with 785 newtons, or 2) it is the scale (along with the solid ground below it) pushing me *up* with 785 newtons, thus preventing me from free falling. Both interpretations are equally valid. There is nothing special about this yet, as long as I see that I am not accelerated. Because then I can easily think that both are correct: gravity is pulling me down, the balance (and the ground) is holding against it. The forces are balanced, so I stand motionless on the spot.

But what, Einstein reasoned, if I can't see if I'm being accelerated, such as when I'm flying around in a spaceship somewhere in space and only feel a force pushing me to the ground. Suppose I can't see if any celestial bodies are nearby. Then I have no way of knowing if the spaceship is just compensating for the gravity induced by a celestial body in the vicinity at a constant speed, or if it is accelerating in weightless space. I can only say: I am accelerated **relative to free fall,** without knowing whether this free fall would just mean crashing onto a celestial body or weightless floating within empty space. The fact that the situation of free fall (and no longer the uniformly moving inertial systems of Newtonian mechanics and SRT) is the reference point for all other forces was elevated by Einstein in 1907 to a principle, the **equivalence principle.**

Gravity is thus downgraded to the status of a **pseudo force.** It's kind of like the centrifugal force: when you ride a merry-go-round, a centripetal force keeps you in a circular path. The centripetal force acts *inward,* toward the center of the carousel, and keeps you from flying straight ahead, away from the carousel. As a carousel rider, however, you instead feel that a force is trying to pull you outward (kind of like when you take a turn fast in a car), a centrifugal force. From this perspective, the centripetal force *compensates for* the centrifugal force, just as the force of the ground on your feet compensates for gravity and keeps you from falling into the bowels of the earth. But you only feel the centrifugal force because you are spinning along with the carousel, so your frame of reference is not an inertial frame, but is rotating. In an inertial frame (and in classical mechanics and SR, these are the "natural" frames), there is only the centripetal force. The centrifugal force is therefore a pseudo force.

In GR, gravity behaves exactly like a centrifugal force: If you feel it as a force (pushing you e.g. against the ground, according to your weight), it is only because you are not in a "natural", i.e. free-falling reference frame. The key thing about gravity is that "free falling" means the same thing to everyone.

All objects, given the same initial position and velocity, have exactly the same trajectory in free fall. This is why Einstein came up with the grandiose idea of also tracing gravity back to the geometry of spacetime, as he had successfully done before with the speed of light.

Since the orbits of free fall are curved in most cases (for a planet, for example, "free fall" means that it moves on an elliptical orbit around the sun), Einstein had to incorporate the concept of **curvature** into this geometry. The math for this is much harder than that of SR, where everything is rectilinear. Einstein was a very intuitive, "physical" thinker. Difficult math was not necessarily his greatest strength. Fortunately, he found that the mathematician Bernhard Riemann had developed the crucial methods, **Riemannian geometry,** to describe curved spaces about 50 years earlier. Nevertheless, it took him 80 years until he was able to present GR in November 1915, a masterpiece of elegance and beauty, born of a few ingenious considerations, brought up with consistent mathematics. Often the way physicists deal with mathematics makes mathematicians shake their heads (Sect. 6.3). However, GR is a counterexample, which meets the highest mathematical demands for precision and thoroughness.

Even better, the theory was relatively quickly confirmed by observations. Its first prediction was that the orbits of planets orbiting their respective suns at relatively close distances deviate slightly from the shape of an ellipse. This deviation ("perihelion rotation") had already been observed for Mercury. In the case of Mercury, much of this deviation was due to the influence of the other planets, but there remained a residue awaiting an explanation, which was then provided by Einstein's theory. The second prediction was the deflection of light rays near heavy bodies (light, though massless, also "falls freely"). This was confirmed in May 1919 during a solar eclipse: Stars whose light passed close to the Sun on its way to us appeared slightly displaced compared to their actual position because the light had made a small arc. This observation was celebrated as the great triumph of GR. The triumphs continued: expansion of the universe, black holes, and gravitational waves are all phenomena described by GR. Cosmic expansion was discovered by Hubble in 1929, we will come back to this.

Black holes are regions of spacetime that are very strongly curved, in such a way that nothing can escape from them, neither matter nor light nor any kind of information. For many decades, black holes were an interesting area of research in theoretical physics. Since the 1990s, however, astronomical observation methods have improved so much that numerous black holes have been detected in the "real world" as well. The best known is the black hole in the center of our Milky Way, which is four million times as "heavy" as our

Sun, i.e. it contains a space-time curvature four million times as strong. Stephen Hawking devoted 40 pages in his *Brief History of Time to* the subject of black holes, which was one of his specialties. So if you want to know more about it, it's best to read the classic.

Expansion of the universe and black holes are also conceivable in Newtonian physics, with slightly modified definitions; in fact, the latter were speculated about already in the eighteenth century, long before Einstein. Gravitational waves, however, have no such counterpart; they only appear in Einstein's theory. They are slight distortions of spacetime that propagate at the speed of light. Their experimental detection in 2016, 100 years after Einstein had predicted them, was therefore once again a very special triumph of GR.

The Curvature of Space-Time

Now how can we visualize and think of a curved spacetime? After all, we have experience with curved surfaces, for example a spherical surface. In order to cover a curved surface with a coordinate system, curved coordinate lines are needed. For a spherical surface, one usually takes a network of longitude and latitude lines. A point on this surface is then defined by its longitude and latitude; these are its coordinates (except for the two poles, where the longitude is not defined). Rules that apply on a flat plane are of no value here, e.g. the sum of angles in a triangle: if you draw a triangle on the surface of a sphere, you will find that the sum of angles is always greater than 180°. This is especially clear if you form a triangle from a quarter of the equator by connecting the ends to the north pole. This triangle has three right angles, so the angle sum is 270°. However, for small triangles that cover only a tiny fraction of the surface of the sphere and are therefore "as good as flat," the sum of angles is almost exactly180° . The **curvature of** the surface is, very roughly speaking, a measure of how much the angular sum changes (away from the default value180°) as you increase the area of the triangle. When the sum of angles becomes larger (as with the surface of a sphere), it is called **positive** curvature; when it becomes smaller, it is called **negative curvature.**

Also length relations like the Pythagorean theorem and its generalization, the cosine theorem, are no longer valid on curved surfaces. In the flat plane, the distance between two points A and B is defined as the length of the straight connecting path AB. Of all the paths, the straight path is the shortest, so we can also say that the distance from A to B is the length of the *shortest* connecting path. On a curved surface, there is usually no straight path, but the second definition can be carried over: the distance is still defined as the shortest connecting path. These paths are called **geodesics.** In the flat plane, the geodesics are the straight lines. On the surface of the sphere, they are the **great circles,** that is, the circles that encompass the full circumference of the sphere. These include the equator and the longitudes. So if there are two points on the equator, it is clear that the shortest connecting path is also along the equator. What path will an airplane take from Frankfurt to San Francisco? If we disregard small detours due to the earth's rotation and wind and weather conditions, it will try to take the shortest connection. To do this, it will determine the great circle on which the two cities lie, and the plane will fly along that great circle, which, amazingly, is over Greenland, much farther north than most would have expected at first. On the flat plane, there is only ever one shortest connecting route. On the curved plane, the same is true in most cases, but in special cases there may be more than one. For the north and south poles, for example, any given longitude is a shortest connecting path.

Just now we characterized curvature as a measure of the change in the sum of angles in a triangle. Another variation is defined in terms of the circumference of circles: In the flat plane, the circumference of a circle is 2π times the radius. That is, if the radius is increased, the circumference increases by the same factor. Now, on the surface of the Earth (as an example of an approximate spherical surface), we take the North Pole as our starting point and move away from it. At regular intervals, we measure the length of the circle that surrounds the North Pole at the same distance, that is, the latitude at which we are standing. We compare this length of the circle with our distance from the North Pole. At the beginning we find that the ratio of the two lengths is 2π. However, the further we get from the North Pole, the greater the deviation from this value. When we reach the equator, the ratio of the circumference of the circle (that is, the length of the equator, the circumference of the earth) and the distance from the north pole (one-fourth of the circumference of the earth) is equal to 4, which is very much less than $2\pi \approx 6, 3$. If we move further away, the latitudes actually get shorter again, until we end up at the South Pole, a latitude of length zero. This gives us the opportunity to understand curvature as a measure of how the ratio of circumference to radius changes (away from the default value 2π) as the latter is increased. If the ratio becomes smaller, as in the case of the surface of a sphere, the curvature is positive; if it becomes larger, it is negative.

A third possible characterization – perhaps the best known – runs through the behavior of parallels. With positive curvature, two geodesics that are parallel at one point approach each other and eventually intersect. This is true, for example, of the longitudes of the Earth's surface: at the equator they run parallel to each other, and at the poles they intersect. In the case of negative curvature, however, the parallels move away from each other and diverge. The behavior of the parallels also explicitly shows that we are dealing with *non-Euclidean* geometry: The axiom of parallels does not hold.

In general, the curvature of surfaces is non-uniform. On the real surface of the earth, for example, there are mountains and valleys, and their surfaces are in turn curved on a smaller scale. In that case, we speak of the curvature at a particular location and can quantify it by measuring circles or triangles in a small neighborhood of that location. The curvature is then a **local** property, so it depends on the exact position.

The characteristics mentioned above can be transferred from curved surfaces to higher dimensions (three-dimensional spaces, four-dimensional spacetimes), even if we then cannot easily visualize this curvature. We should think of angle and length relations, or of the behaviour of parallels, which deviate from those in Euclidean space or those in the space-time of SR. These deviations lead to the fact that the shortest connection of two points is no longer a straight line. Riemannian geometry, which is the mathematical basis of GR and which Einstein had to learn so laboriously between 1907 and 1915, describes how the curvature and the shape of the geodesics arise from the *metric*, i.e. from a generalization of the relation in Eq. 7.1 or in Eq. 7.2. The metric is now also no longer the same in the whole spacetime, but depends on the position and the time and can thus be understood as a *field*. Therefore, GR is a *field theory* which describes the behaviour of this field.

To reduce misunderstandings (they can never be completely avoided in popular scientific presentations), I would like to explain the concept of **embedding** before we return from mathematics to physics. The two-dimensional surface of a sphere is **embedded** in a three-dimensional space. The name says it all: we think of it as the surface of a three-dimensional structure, a sphere. It divides the space into an inside and an outside, so to speak. But we can also think of this surface without the embedding (although not really visualize it), as a surface with certain, abstractly defined distances and curvature properties, without thinking of the third dimension, the inside and outside, in which it is embedded.

This can be explained more simply using the circle line: We know the circle as a shape in two dimensions. But the circle itself is one-dimensional, a line. Can we also define it as a purely one-dimensional structure without adding a second dimension? Yes, we can. We simply say: A circular line is a line of finite length whose starting and ending points we *identify with* each other, i.e., we simply say by definition that the ending point E is "the same" point as the starting point A. It then follows that there is actually no starting and ending point at all: If we move to the right beyond E, we come back in on the left at

A, precisely because we have defined the two points to be identical. By this definition, the two-dimensional circle we know is an *embedding of* this abstract circular line. To *realize* the identity of A and E, we bend the line around in the direction of a second dimension, so that we can "glue" A and E together. If we do the bending perfectly evenly, the result is the two-dimensional circle, the circular line is *embedded in* two dimensions.

The big brother of the circle line is the torus. The torus can be defined abstractly as a rectangle, where the upper side is *identified* with the lower side and the left side with the right side. You know this from some old computer games: An object that moves out of the screen on the right comes back in on the left, and vice versa; or an object that moves out of the screen at the top comes back in at the bottom. Defined in this way, the torus is a flat, two-dimensional structure. Much better known from the literature is the *embedded torus*, i.e., the torus is *realized* in three-dimensional space by bending down the top side and gluing it to the bottom, and then bending the resulting "tube" again to form a hoop so that the right side can also be glued to the left. The resulting embedded torus is shaped like a ring, bagel, or doughnut, and this is how the layman generally imagines it. Therefore, in popular scientific literature, the torus is often characterized as a structure with exactly one hole. However, the abstract torus has no hole at all; the hole comes about only because of the embedding.[2]

For GR, it is important not to think of curved spacetime as coming from embedding. That is, the curvature does not come from the fact that spacetime is bent, for example, in the direction of an additional fifth dimension. Many curved spaces are not easily embedded in a flat higher-dimensional space. Already the negatively curved equivalent of the spherical surface, the "hyperbolic plane", cannot be realized in a flat three-dimensional space.

Einstein's Field Equations

The core of GR are **Einstein's field equations.** On the left side of these equations are purely geometrical expressions, which are associated with the curvature of space-time. On the right side are expressions describing the matter moving in this spacetime, in particular its energy and momentum. Here, mass is a part of energy, as known from SR. This can be interpreted in such a way that the matter *causes* the curvature.

---

[2] However, mathematicians have expanded the concept of "hole" so that even the abstract torus has a hole. It's all a matter of definition.

A free-falling object, i.e. an object that is not affected by any force other than gravity, moves through spacetime on a geodesic. The geodesics in turn are determined by geometry, in particular by curvature. The influence is therefore reciprocal: matter determines the curvature of spacetime and this in turn determines the motion of matter.

Further above I said, a geodesic is the shortest connection between two points. But this is only true for points whose distance is space-like. For time-like distances, it is the *longest* connection between the two points. This is because for time-like distances, any detour makes the distance shorter, not longer. Let us recall Erwin and Otto. Erwin moved in a straight line from A to C, covering a spacetime distance of ten units. Otto, on the other hand, moved in a zigzag, first from A to B, then back to C, a detour, but he covered only six units; he aged less (Fig. 7.5). The shortest connection between two events separated in time is always zero: a ray of light always travels the spacetime distance zero. To travel the shortest distance from A to C, you ride a light ray away from Erwin (i.e., even much faster than Otto) and then switch to a light ray in the opposite direction, back to Erwin. The spacetime distance traveled is zero. Therefore, the straight path Erwin takes is the *longest* path. In a curved spacetime, it is not a straight path, but a geodesic.

A freely falling object thus moves on a geodesic. The Sun produces a rather weak curvature of spacetime in its vicinity, so the geodesics deviate only slightly from straight lines. Free falling means for planets that they move on an ellipse orbit around the sun. But an ellipse now looks like a very sharp deviation from a straight line. How does that fit together? Well, you have to look at the orbit in spacetime, not the orbit in space. As the Earth travels about 900 million kilometers in its orbit around the Sun, it also moves 1 year into the future. In space-time, this is not an ellipse, but a helix. One year corresponds to a distance of exactly one light (the conversion factor from the time unit year to the length unit light year is exactly 1:1, because the speed of light is exactly 1), which is 9.5 trillion kilometers, or 10,000 times the circumference of the ellipse. So the helix is really almost straight. If you draw the helix (with a single turn) on a piece of paper 10 meters long, the radius of the turn is just 0.15 mm. This is less than the thickness of a normal pencil line. So the line looks exactly like a straight line on the paper. The curvature is very small indeed.

One of the most remarkable and tragic stories related to GR is that of Karl Schwarzschild. When Einstein published his field equations in November 1915, he saw that they were very beautiful, but also very complicated. He assumed that in almost all cases they could only be solved by approximation methods. (Solving here means, for certain configurations of matter, determining the metric as a function of location and time in such a way that the equations are satisfied; everything else then follows from the metric, in particular the geodesics and hence the trajectories of the freely falling bodies). He had also found his result for the perihelion rotation of Mercury with such an approximation method. He was all the more astonished when, a few weeks later, he received a letter from Schwarzschild in which the latter presented the *exact* solution from which the perihelion rotation followed, without any approximations. *"The following lines thus lead to let Hr. Einstein's result shine in increased purity,"* Schwarzschild wrote in the paper, which was published shortly thereafter, in early 1916. This solution of Einstein's field equations, the **Schwarzschild solution,** is still the most important solution known today. It describes the geometry of spacetime in the vicinity of a point-like or spherically symmetric mass distribution such as the Sun, a planet, or a black hole. It also shows that Newtonian gravity predicts almost the same trajectories as GR in normal cases (i.e. at sufficient distance from the central mass, e.g. the Sun), but that these deviations become larger as one approaches the central mass or if this mass is very large.

Schwarzschild, the director of the Potsdam Astrophysical Observatory, had volunteered to join the army in the widespread enthusiasm to take part in the First World War. After some time on the Eastern Front, he fell seriously ill and was hospitalized. Both at the front and in hospital he kept himself constantly up to date on the latest physical and astronomical research. Some of his most important papers date from this period. He received Einstein's paper on ART while still in hospital in November 1915. Within a few days, from his hospital bed, he found his famous solution. Schwarzschild died a few months later as a result of his illness, at the intellectual height of his creative powers.

Cosmological Solutions

Another important class of solutions are the **cosmological solutions.** These assume that the universe looks more or less the same, broadly speaking, in all directions of the sky. By this is meant that the stars and galaxies in the universe play a role similar to that of the mountains and valleys on Earth: when zoomed in close, they form widely varying regions of different curvature. But from a greater distance, the Earth is an almost completely symmetrical sphere; the mountains and valleys are small bumps that only slightly disturb the uniformity on a large scale. The cosmological solutions thus describe a uniform universe at large, ignoring the small spacetime mountains and valleys that disturb this uniformity when zoomed in.

These cosmological solutions, surprisingly, have some properties that undo some of the peculiarities of SR. Therefore, cosmology is in some respects closer to Newtonian physics than to SR. For simplicity, let us consider only cosmological solutions with **big bang**, i.e. with a temporal beginning in which space had no extension at all (everything was crowded into one point). From that moment, the universe began to expand, that is, things in it began to move away from each other. It seems that such a solution to Einstein's field equations describes the universe we live in quite well (more on this in Sect. 7.9).

If there is such a temporal beginning, then we can again define an absolute time (which was not possible in SR): The absolute time of an event A is defined by the distance between A and the big bang. The distance can again be defined as the *longest* connecting line (it is clearly time-like). So we can say A takes place so many years after the big bang. In this way we can also neatly separate time and space, e.g. we can say "the universe 3 min after the big bang", and by this we mean the **space** 3 min after the big bang, and this is the set of all events (spacetime points) that take place 3 min after the big bang. This is the only reason why we can speak of an "age of the universe".

Similarly, we can again find an absolute system of rest: We simply define that whoever is moving along the longest connecting line between the Big Bang and an event A is at rest (this works because there is in fact only one such longest connecting line between the Big Bang and A), i.e., we define the points on such a line as "the same point in space" (because whoever is at rest always stays at the same point in space). One must realize that this is a peculiarity. In SR all uniformly moving reference frames are equal, you can't say of any in an absolute sense that it is at rest, because each is moving relative to all the others. Thus, in SR, one also cannot say, "I am at the same point in space as before", because the same point in space is not defined at all over time. In the cosmological solutions of GR all this is suddenly possible again.

Also, the limit of the speed of light no longer applies. The expansion is such that objects which are at rest in the cosmological sense just defined are nevertheless moving away from each other. An expanding universe means that distances between objects at rest are constantly increasing. This kind of expansion comes about mathematically through the properties of the cosmological metric. If we want to understand it in everyday language, I think the best idea we can come up with is that new space is created between two points in space without the points actually moving away from each other. The farther apart two points are, the faster their distance increases. Since two points in space can be arbitrarily far apart – there are cosmological solutions in which space is infinite – there is no limit to this increase in distance. We can express the increase in distance in terms of a velocity. It is not a velocity in the proper sense, since the objects are at rest. Nevertheless, "increase of distance per second" has the form of a velocity. As said before, this speed can be arbitrarily large, in particular larger than the speed of light.

Also the curvature of space is unambiguously defined in such a solution, not only the curvature of spacetime, and this is because we could separate space and time so neatly and make them absolute. If space is positively curved, it behaves much like a spherical surface, only with one dimension more (the spherical surface has two dimensions, space three). In particular, the universe is then finite, and if we always fly straight, we will eventually return to the starting point, as if traveling around the Earth. If space is flat or negatively curved, then things are a bit harder to decide. We'll come back to this in Sect. 7.9.

An important property of GR is that the law of conservation of energy is generally no longer valid there. This can be seen particularly clearly in the cosmological solutions: A light ray is pulled apart by the expansion, thereby the wavelength increases accordingly. However, the energy of light is antiproportionally related to its wavelength. Therefore, during the expansion, energy is lost without being transferred into another form.

In SR and GR space and time form a unity, the spacetime. Time behaves like a fourth dimension of space. The only difference is the minus sign in the Pythagorean theorem. Why then do we perceive space and time so differently? Does this difference, so radically perceived, really follow from this one minus sign alone? To what extent can we trust our perception on this point at all? This question arises in particular because we associate certain ontological ideas with time: The past *no longer* exists, the future does *not yet* exist. But what sense do these propositions make when we realize that space and time constitute a common whole, a spacetime given quasi *at once*, from the temporal beginning to the end, a so-called **block universe?** We will come back to this.

Normally, by "universe" we mean the totality of space, not the totality of spacetime. When we speak of the age of the universe, we mean *space now*, which is a certain distance from the Big Bang. But that we can speak of *space now* so unambiguously at all is only possible because of a peculiarity of cosmological solutions; it is not something that has any meaning at the level of the fundamental laws of SR or GR. It would be better to mean spacetime itself when we speak of the universe. But spacetime has no age, because it contains all of time from beginning to end.

7.5   Statistical Mechanics

As early as ancient Greece, some philosophers were certain that all matter was composed of **atoms**, small indivisible particles of which there were only a few different varieties (four or five according to the view of the time), and from the interaction of which all the diversity of the observable world emerged. After the Dark Ages were overcome, some natural scientists of modern times took up the hypothesis again. But at first it was only a metaphysical speculation that could not be proved. With the tremendous advances in physics and chemistry in the nineteenth century, however, the hypothesis became more and more plausible. In chemistry, the existence of chemical elements crystallized, substances that could form chemical compounds with each other but were not themselves compounds of other substances. This could be explained by the fact that each element consists of a certain type of atom, whereas chemical compounds consist of molecules, i.e. particles that are firmly composed of several atoms. Since there are quite a lot of elements, about 100, this had increased the number of atomic species by about 20 times compared to ancient times.

With the help of atoms, the three aggregate states – solid, liquid, gaseous – can also be understood. In the solid, the atoms (or molecules, in the case of compounds) are in fixed positions. They can shake back and forth a bit, but they can't change their position relative to each other, much like spectators in a football stadium. Let's stick with this analogy for a moment. When the football game is over, the mass of fans melts into a liquid: they leave their fixed seats and stream toward the exit. They still move close together, but their relative positions to each other are no longer rigid. The volume of the crowd remains the same, it is not compressible (because everyone is already moving close together), but its shape is changeable. Behind the exit, the spectators evaporate into a gas: they lose physical contact, expand (i.e. not the individual person expands, but the totality of the crowd), flowing apart in all directions.

In the eighteenth century, the steam engine emerged as one of the most important tools of industry. The steam engine is an example of a **heat engine** in which, as the name suggests, heat is used to exert a force and thus perform mechanical work. This is accomplished by means of a gas, in this case water vapor, which when heated tends to expand its volume and exerts a pressure on its bounding surfaces. One of these boundary surfaces is a movable piston, which is thereby pressed outwards (as seen from the gas). Thus it is already clear that **heat is** to be regarded as a **form of energy** which is converted into kinetic energy by the heat engine. The **first law of thermodynamics** is nothing else than the classical law of conservation of energy extended by the concept of heat. Heat, for its part, is typically generated in heat engines by chemical reactions, such as the combustion of a fossil fuel. Heat thus serves as a link in the conversion of chemical binding energy into classical kinetic energy.

The reverse conversion, namely from kinetic energy to heat, happens through **friction.** Rub your palms together and you'll know what I mean. Friction slows down everything around us, because it occurs wherever two boundary surfaces move against each other. Because of friction, a vehicle must always be expending energy to keep moving, because friction is always draining energy from that movement and converting it into heat. Thus friction is responsible for the greatest misunderstanding in the history of physics: most scientists before Newton thought that the concept of force was to be defined in terms of *motion.* You needed a force to set a thing in motion and to keep it in motion. Without force, they said, everything stops. That is why they misunderstood the motion of the moon and the planets. They thought that a force had to push these celestial bodies permanently so that they would not stand still, and they wondered how this could happen. Only Newton cleared up the misunderstanding: It is precisely *without the* action of a force that things remain in uniform motion. The fact that this is not the case with sliding and rolling, pulling and pushing on earth, is solely due to the fact that friction is a force that constantly slows everything down and must therefore be overcome by a counterforce. The celestial bodies move in a vacuum, there is no friction acting on them, and therefore they do not need to be constantly pushed.

Thermodynamics is the technical term for heat theory. It deals among other things

- with the relationship between temperature, pressure and density of gases,
- with the amount of heat (i.e. amount of energy) contained in a given amount of a given substance at a given temperature,
- with the dynamic compensation of temperature differences (heat conduction),
- with heat engines and related cyclic processes,
- with connections between heat and magnetism and
- with certain aspects of chemical reactions.

In thermodynamics, heat is first of all a physical quantity that is not explained further, so its exact meaning remains unclear at first. In the second half of the nineteenth century, however, it was possible, especially by Maxwell, Boltzmann and Gibbs, to give thermodynamics a deeper theoretical basis through

statistical mechanics. Here, heat and all related phenomena are explained from the statistical behavior of the smallest particles, atoms or molecules. So the atomic hypothesis is quite crucial to understanding these things.

Only at the beginning of twentieth century, atoms could be detected directly by experiments. Until then, still some physicists insisted, that although everything could be explained very well by pretending everything to exist of atoms, atoms are still pure theoretical constructs, which better shouldn't be assumed to exist in reality. Especially the physicist and science theorist Ernst Mach (who with some other thoughts inspired Einstein to his GR) held this view.

In the twentieth century, things developed rapidly. It quickly became clear that what were called atoms were not indivisible at all, but consisted of electrons, protons and neutrons. The large number of different kinds of atoms only came from the fact that these particles could be combined in different quantities. Basically, the word "atom" in its current meaning is out of place, because only the elementary particles are indivisible.

So what does statistical mechanics say? It explains heat as the undirected motion of the smallest particles (atoms or molecules) of a substance. It differs from mechanical kinetic energy in that it is *undirected*, i.e. each particle moves on its own and independently of the others in a confined space, so that the totality of these movements cannot be seen from the outside. How much freedom of movement the particles have depends, among other things, on their state of aggregation. In a solid they can only tremble back and forth at their fixed positions, in a liquid they can also turn and move past each other, in a gas they can even fly quite a distance until they collide with the next particle. Due to the constant interactions with each other, the different directions of motion balance each other out, energy and momentum are transferred back and forth, a very specific statistical distribution of velocities and energies is created. The average value of this distribution corresponds to a certain **temperature.** If heat is added to an object, e.g. by friction, radiation or electric current, then the energy is distributed evenly by the constant collisions and other interactions of the particles and increases the statistical mean value of the energy per particle, i.e. the temperature increases.

At **absolute zero,** at about -273 °C, the energy per particle is zero. It can't get any colder than that. From a physics point of view, it makes sense to choose the units of temperature so that at absolute zero the temperature value is zero, rather than -273 °C. This is why physics uses the Kelvin scale, which uses the same degree intervals as the Celsius scale, but is shifted to the "left" by -273 °C so that it has a value of zero at absolute zero.

Statistical mechanics distinguishes between the **macrostate** and **microstate of** a physical system. The macrostate is what one can see or measure of it as a macroscopic observer: external form, chemical composition, mass, volume, pressure, temperature, magnetization, etc.. The microstate, on the other hand, contains the exact state of each particle, i.e., its respective position, velocity, and possibly other details such as rotation and magnetic moment. The macrostate depends only on statistical summaries (averages) of the microstate. We do not need to know the microstate to do thermodynamics with the macrostate.

A macrostate represents an **ensemble** of microstates: The statistical averages that characterize the macrostate may arise from a huge number of different microstates. Which of these is realized, we cannot possibly know. From the temperature, we only get the *mean of* the velocities, but not the actual velocity of each particle. In theory, therefore, the macrostate is conceived as the *totality* of microstates for which the mean values of various physical quantities are compatible with the values of the macrostate. This totality is here meant by the term ensemble. It is a useful term only for theoretical purposes, because we mostly assume that only exactly one of these microstates is realized in nature, we just do not know which one because we cannot measure all particles at once. The ensemble term also expresses an affiliation: A microstate X *belongs to the* macrostate A if the statistical mean values in X are compatible with A.

Entropy

One of the most important concepts in statistical mechanics is **entropy.** The entropy of a macrostate is defined in terms of the amount of information needed to uniquely characterize a microstate within that macrostate. In popular scientific accounts, entropy is often characterized as a measure of the "disorder" of a state. The connection becomes clear with an example. Suppose I am doing statistical mechanics with my books. Each book represents a "particle" that is somewhere in my living room. At first, everything is very tidy. The macro state is "all books are on the shelf." Then I can describe the state of each book (i.e., the micro state) fairly easily: Each book is closed and standing upright; I can easily list the order from left to right and top to bottom. Then an acquaintance comes to visit with his 3-year-old son. The latter pulls out books at random, lays them on the floor, flips through them, takes out more books, spreads them all over the room. The resulting macro-state might be described in everyday language as messy: "The books are scattered criss-cross around the room." In terms of entropy, it means that I have to give someone who can't see the room a lot of information to describe the microstate, the state of each book: The exact coordinates in space where it is; which way it is turned; whether it is open and, if so, on which page. Much more information is needed than in the first, "ordered" macro state. So the entropy is higher. The fact that much more information is needed to describe the microstate is again because in the "disordered" case there are many more possibilities of how the books can be positioned, oriented, and open. The information I give must distinguish between these possibilities and express which *of* them is realized.

Since the concept of entropy is essentially related to information, it is not surprising that it also proves useful in computer science, e.g. in the field of data compression. Suppose you have a file of size 10 MB. This file then consists of about 80 million bits (1 byte is 8 bits, 1 MB is 1024 by 1024 bytes), where each bit has a value of 1 or 0. If the ones and zeros are completely randomly distributed, then you really need to specify their exact sequence to uniquely describe the contents of the file. Entropy is at a maximum; you can't compress the file any further. The opposite extreme occurs when the file consists only of ones. Then, to describe the contents, you don't need to say "one" 80 million times; it's enough to say, "The file contains 80 million ones in a row." There is maximum "order", i.e., entropy is minimal, and the file compresses perfectly. For almost all realistic files, the truth lies in the middle, i.e., there are certain patterns that repeat frequently (e.g., recurring words in a text file), or uniform sections (e.g., blue sky in an image file). Each of these patterns makes it possible to summarize some of the file's contents, so you need a little less information, so in the end the entropy is somewhere between the minimum and maximum values. The smaller the entropy, the more the file can be compressed.

The direct relationship between entropy and heat is that a physical system has more entropy when it is warmer. Take, for example, a gas enclosed in a fixed volume. Then the microstate is characterized by the positions and velocities of the particles. If the gas consists of molecules (and not individual atoms), their rotations and vibrations are also relevant. Suppose the gas is first cold and then heated. How does the amount of information needed to describe the microstate change? Nothing changes for the positions of the particles, since we assume that the volume is not changed. However, after heating, the particles are on average faster than before. The velocities follow a certain statistical distribution, some are still very slow, some are much faster than average. Crucially, the distribution is *wider* after warming than before. You can say there are more different velocities available to the particles. So there are more possibilities to distinguish between. (In principle, at least in classical physics, i.e. without taking QM into account, there are infinitely many possibilities if we want to fix a velocity absolutely precisely, i.e. to infinitely many decimal places. But as everywhere else in physics, we assume a finite resolving power, i.e., we want to fix the velocity only to a certain accuracy, say, to one millimeter per second. Then we can count off the possibilities. If, on the other hand, we take QM into account, the finiteness – and hence countability – of the states follows directly from the theory). The same is true for the rotations and oscillations of the molecules, they also have more possibilities available after heating. That is, the entropy has increased.

Entropy plays a major role in chemical reactions. Suppose a mixture of substances can undergo two different reactions, one leading to macrostate A, the other leading to macrostate B. At the particle level, the reactions are described by QM, which tells us the probability of microstates being reached in A or B. Let's say it turns out that a microstate in B is on average about a million times more likely to be reached than a microstate in A. But A has a higher entropy, such that there are a billion times as many microstates available for A as there are for B. The two pieces of information together result in the reaction primarily running towards A because the higher number of microstates outweighs the lower probability per microstate by a factor of a thousand.

The Second Law and the Direction of Time

The second law of thermodynamics states that the entropy of an isolated system (i.e. a system that does not exchange particles or energy with its environment) never decreases. There are **reversible processes** in which entropy remains the same, and **irreversible processes** in which it increases. A macrostate is at **equilibrium** when its entropy is at a maximum, i.e., when there is no way to increase its entropy further (unless you remove its isolation and supply energy to it from the outside, e.g., by heating it up). It follows from the second law that isolated systems will seek such an equilibrium in the long run. Since entropy can only increase, such a system cannot move away from equilibrium. With each irreversible process, however, it comes a little closer to equilibrium.

For many systems, it is relatively easy to calculate what a state of equilibrium looks like. It turns out in these cases that at equilibrium everything is fairly evenly distributed. At every point in the system, there is the same temperature, the same density, the same pressure; each type of particle is evenly distributed over space, and the different types of particles are in certain fixed relationships to each other, determined by the details of their possible reactions.

The latter circumstance means that the equilibrium defined via entropy also includes **chemical equilibrium in** particular. In chemical reactions and their equivalents in particle physics, certain types of particles are transformed into other types of particles. Reactions can always proceed in both directions, i.e. for every *reaction* there is also a *backward reaction*. In chemical equilibrium, the ratios between the numbers of the respective particles are such that the reaction takes place in both directions with exactly the same frequency, so that nothing changes in the ratios.

Such a state of equilibrium actually sounds pretty ordered. It may therefore be surprising that it is accompanied by maximum entropy, which is after all a measure of "disorder". This shows that "disorder" as a metaphor for entropy has limited applicability after all. It is also interesting that the macrostate with maximum entropy is particularly simple, i.e. one needs very *little* information to describe it (temperature, density, etc., which are the same everywhere). In return, one needs particularly *much* information to specify a microstate in it. In the case of a non-equilibrium state, you need more information for the macrostate ("it's a bit warmer here, a bit colder there"), but less for the microstate.

If we are dealing with a system of rigid bodies, their particles cannot mix with each other, the bodies remain separate. In this case, the equilibrium of particle distribution cannot be aimed at; the associated irreversible processes, which would cause the bodies to mix, do not take place (one would have to melt the bodies for this and let them flow into each other). But the system is still evolving towards an equilibrium of the temperature, i.e. temperature differences are balanced in the long term.

The equilibrium distribution of temperature leads to the conclusion that temperature differences only ever decrease by themselves, not increase. Heat does not flow by itself from a cold to a warmer subsystem, but only vice versa (this is why our refrigerator needs constant energy supply to prevent gradual adjustment to the outside temperature). This statement is itself often referred to as the second law of thermodynamics. Historically, the second law has gone through many different formulations and has always been understood slightly differently. At first, it was a purely experiential statement about how certain thermodynamic processes work, which ones can be reversed and which ones cannot, etc. The reduction to the concept of entropy came only later. But even with the theoretical foundation of statistical mechanics, things are not so clear. To this day, it is certainly the most hotly debated piece of nineteenth century physics. The second law is not a real fundamental law of physics like

Maxwell's equations, for example, just as statistical mechanics in general only draws *conclusions* from the fundamental laws with the help of statistical (i.e. mathematical) methods.

So, how does the second law follow from the basic laws of physics? The shocking answer is: not at all! On the contrary, it follows directly from the fundamental laws that the second law is false. This is because of the time-reversal symmetry of the laws of nature. This symmetry states that any temporal evolution can just as well proceed forward as backward. More specifically, if a system evolves from microstate A to microstate B within a given time period, then it can also evolve in the opposite direction, from B to A, within the same time period. To do this, one only needs to rotate the velocities of all particles in state B in the opposite direction, then the whole event runs backwards and ends at A.

You can easily imagine this using our planetary system as an example. Let's say a space probe films the solar system from the outside for a few years. Let's run the film forwards and backwards. Both versions look equally realistic, the only difference being that all the motion is in the reverse direction. And in fact, the backwards version is just as good a solution to the laws of gravity (whether you go with Newtonian or Einsteinian theory) as the forward version. The same is true for all other known laws of nature. (For some particular laws of nature, namely those of the weak nuclear force, right and left must additionally be swapped and particles replaced by their antiparticles – more on the concept of antimatter later – but this has qualitatively no influence on our discussion). For GR, this means that any spacetime can also be "turned upside down", i.e. the direction of time can simply be reversed, future and past can be swapped.

It follows for entropy that for every process in which entropy increases, there is also the opposite process in which it decreases. The second law is therefore wrong. It does not even hold statistically, i.e., it is not even the case that an increase in entropy is *more likely to* occur than a decrease. For there is a precise one-to-one relationship between theoretically possible systems in which it increases and those in which it decreases.

Now how does that fit together? We can clearly see that many processes only occur in one direction. A glass falls to the floor and shatters into shards, but no one has ever observed that shards spontaneously spring up from the floor and reassemble themselves into a glass – nor that a steak in a hot pan changes from the fried to the raw state. The fact that we come into the world as old men and die as infants only exists in the movie about Benjamin Button. Imagine what it would look and feel like if our digestion and food intake went backwards. It all seems totally absurd. The second law and the irreversibility of many processes are clearly correct.

The solution to the problem, interestingly enough, is not in the laws of nature, but in the special *solution of* the laws of nature, which represents our world. The second law is directly connected with the problem of time direction and causality. At the level of microstates, that is, particles, no direction of time is defined. Since everything can just as easily proceed forward as backward, and any spacetime can simply be turned upside down, there is no distinction between before and after. Since cause and effect are arbitrarily interchangeable in this way, there is also no causality. It is a great mistake to think that deterministic theories explain causal relations to us. On the contrary, they completely dissolve the notion of causality, because one can calculate the state at time t_1 just as well from the state at time t_2 as vice versa. Before or after does not matter.

For a macroscopic observer, who only knows the macrostates, the matter is different. Let's take a closer look at the example with the glass falling to the ground. When it hits the ground, the entropy increases significantly: first, the glass breaks apart, the shards spread on the ground "in disorder"; second, the kinetic energy E of the falling glass is transformed into the amount of heat Q, which is distributed through the ground to the entire planet. The large entropy difference between macrostate A (whole glass falling) and macrostate B (shards lying on the ground, planet Earth having absorbed quantity of heat Q) means that the number N_A of different microstates belonging to macrostate A is much smaller than the number N_B of microstates belonging to macrostate B. Suppose the ground is hard enough at the point of impact and the glass is fast enough that we can predict with certainty that the glass will shatter. That is, if we assume A (i.e., start with the falling glass), it develops into B with one hundred percent probability. The microstate is irrelevant; the event can be narrated and understood on the basis of the macrostates alone.

Due to the time reversal symmetry of the microscopic physical laws, the entire event can be reversed in time. That is, it can happen that the macrostate B (shards lie on the ground) changes into the macrostate A′(whole glass rises upwards, planet earth has given off quantity of heat Q). Imagine the backward process in detail. There were N_A microstates of A where A leads to B, so there are also N_A microstates of B where B leads to A′. But the total number N_B of microstates in B is, as I said, a huge factor greater than that. Therefore, the N_A microstates that cause the shards to jump up and form a glass form only a tiny fraction in the totality of all microstates in B (Fig. 7.7).

This is easy to see: a large number of particles of the ground must cooperate quasi "conspiratorially" in order to jointly give the lying shards a push in

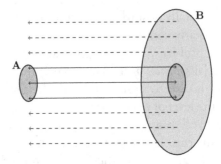

Fig. 7.7 A macrostate A evolves to a macrostate B with higher entropy. Higher entropy means that many more possible microstates belong to B than to A. Because of time reversal invariance, all processes can also run backwards on the level of microstates. So there are just as many microstates running from A to B as from B to A. However, these make up only a tiny fraction of *all* microstates in B. If B is given, the probability that B will evolve to A is therefore virtually zero

exactly the right direction; in addition, the particles at the edges of the shards must cooperate "conspiratorially" in order to stick together with the edges of the other shards so precisely that a coherent glass is formed without visible cracks and scratches. For the first part, that is, the thrust, the heat of the ground must spontaneously contract in the points under the shards, in such a way that the particles which are under the shards all make a little movement together in the same direction at the same time. This movement gives the shards the push upward. It is the exact reverse of the motion caused in the first case (glass falling and shattering) by the shards hitting the ground.

This makes it clear what the difference is between the two directions of the process: In the first case, what happens can be understood from the macrostate, the glass falls to the ground and must shatter. In the second case, what is happening can only be understood from the microstates, that is, you have to know the exact motions of all the particles to know why something is happening. These microstates are, in relation to the microstate B', so improbable, so "conspiratorial", that they seem completely unnatural to us. Moreover, we cannot observe all the particles individually; that is, if it should ever happen that broken pieces suddenly jump up and form a glass, it will seem to us like a miracle, like an "intervention of God". The causal chain is only recognizable to us in the one time direction, namely in the first case, where the glass falls to the ground and entropy increases.

This can be generalized. Suppose you are handed a piece of spacetime, framed between two points in time t_1 and t_2. You don't know which is up and which is down, that is, whether t_1 is earlier or later than t_2. You only know that the piece of spacetime you hold in your hands represents the history of the universe between those two points in time. At the microscopic level, it makes no difference which way around the story goes. You can recalculate how the microstate at the point in time t_1 evolves to the microstate at the point in time t_2, or vice versa. The order doesn't matter. But now you're supposed to explain to someone what happens between those two points in time, without demonstrating the whole complicated calculus. And there you are relieved to find that the entropy at the t_2 is much smaller than at t_1. Because then you can *define* that t_2 represents the *beginning* and t_1 the *end*. For in this direction, events can be told naturally in terms of macro-states (falling glasses; frying steaks; infants becoming old men). Cause and effect can be understood at this level. In the opposite direction, on the other hand, you would have to bring each particle into play and show how it interacts with the others to produce a particular effect, as with the shards spontaneously leaping up. **Only from a macroscopic point of view does the direction of time make a difference. Only from a macroscopic point of view is there cause and effect in a definite order, i.e. causality.**

The Big Bang could represent the beginning or the end of the universe, on a microscopic level it doesn't matter. But near the Big Bang, entropy is very low and increases the further away you get. That's why we see the big bang as the beginning, not the end, and can understand and tell the story of the universe much better that way around. We say the universe has evolved from the Big Bang to the present, and present events were caused by past ones. We remember the past, the future is uncertain. Viewed from the opposite direction of time, in which the Big Bang is the end, we foresee the future as we move toward our birth, and know nothing about the past. This view contradicts our experience and understanding of causality, but from a microscopic point of view it is as correct as the other.

The logic that leads to the second law of thermodynamics is thus the following: We live in a universe that is very asymmetric in time, in such a way that entropy is consistently increasing in one direction. This direction is, for us as macroscopic beings, the natural direction of time, that is, the direction in which we distinguish past and future and in which we understand causal relationships. That entropy only increases, but does not decrease, is simply a paraphrase of the fact that we have naturally *chosen* the direction of time such that it increases.

Why the universe exhibits such a strong temporal asymmetry as far as its entropy is concerned is still an unanswered question. The currently known that the laws of nature *allow for* such an asymmetry, but do not seem to favor it. However, it may well be that there are as yet unknown laws of nature that shed new light on this behavior. For a more detailed treatment of this multi-faceted topic, see the book *From Eternity to Here* (Carroll, 2010).

In most problems of statistical mechanics (or thermodynamics), a time direction is already assumed, and initial conditions are given in the form of macrostates. In this case, the second law applies *statistically:* Apart from a few "conspiratorial" microstates, these states will almost always evolve towards macrostates with equal or higher entropy. If, on the other hand, one were to specify final states instead of initial states and calculate them back, then one would obtain the opposite of the second law: the initial state would almost always have the same or higher entropy than the final state. So you would get a result that does not seem to agree with reality. But since one has already assumed the direction of time, one will not take this second path, but will always calculate (let evolve) everything from the initial state, not from the final state. The macrostate at the end then no longer contains enough information to recover the initial state from it.

7.6 Quantum Mechanics

You probably think SR, GR and the thing about time reversal are already quite strange theories. But all this is nothing compared to QM. No other theory can match QM for strangeness. It solves thousands of problems, makes exact predictions about the behaviour of microscopic and partly also macroscopic objects. But for common sense it does not make the slightest sense. That's why many physicists were reluctant to accept it. SR and GR were a triumph of mind. Based on clear, unambiguous considerations and physical principles, the theories were derived and passed experimental tests. Newton's mechanics and Maxwell's theory of electromagnetism had already presented themselves in a similarly aesthetic manner. Statistical mechanics is also based on clear mathematical principles. Nothing of the sort with QM. No nice principle was behind it, no clear reasoning. One can say nature has violently forced us to put all the ghastly puzzle pieces together into an even ghastlier whole that works so incredibly well that we can never get rid of it. There is only one name associated with each of many of the other theories: Newtonian mechanics, Maxwellian electrodynamics, Einstein's GR and SR. For statistical mechanics, Maxwell, Boltzmann and Gibbs did the main work. For QM, however, it took quite a few physicists to put the pieces of the puzzle together: Planck, Einstein, Bohr, Heisenberg, Schrödinger, Dirac, Pauli, Born, de Broglie, Jordan, von Neumann, Wigner, and later Bell and Everett.

QM was a watershed in the history of the natural sciences. Many compare it with the Copernican revolution or with Darwin's theory of evolution. For the science theorist Thomas Kuhn, it is a typical example of a scientific revolution and a paradigm shift. In physics, it marks a division: between classical physics and quantum physics.

The strange contrast of QM is that, on the one hand, it is a very fundamental and universal theory that provides exact mathematical descriptions and makes predictions in a huge variety of situations, but on the other hand, it provides no real explanatory framework, no world view. You can't tell from it what it actually "means" and what "things" it is about. Until today physicists and philosophers try to decode its meaning, its sense. There are almost as many interpretations of QM as there are physicists who deal with it. And, as is so often the case, everyone is convinced that their interpretation is the only correct one (this is true of me, of course). However, it is only a small minority of physicists who deal with these questions, because it is not particularly conducive to a scientific career, since one does not come to any unambiguous conclusion (at most unambiguous for oneself) and exposes oneself to ideological trench warfare. In other words, one is not really doing natural science, but metaphysics. However, every now and then there is a truly scientific result that rules out certain interpretations of QM as untenable.

Interestingly, the question of interpretation is completely irrelevant to the scientific content of QM. One can wonderfully calculate with it and test its predictions in experiments without giving the slightest thought to the meaning. This contrast between scientific exactness and simultaneous obscurity of meaning is what makes QM so unique. It is also the basis for the proverbial "shut up and calculate" mentality that has dominated physics since the middle of the twentieth century (further reinforced by the properties of QFT; Sect. 7.7): Just calculate and don't worry so much.

At the beginning of the twentieth century it was clear that atoms consist of a very small atomic nucleus and a shell in which the electrons scurry around. The crucial question was how such an atom could be stable: If the electrons don't move, they crash into the nucleus because of electrical attraction. But if they move around the nucleus, according to Maxwell's theory, they must emit radiation, thus losing energy and also crashing into the nucleus. Somehow the electrons must move and not move at the same time for the atom to persist. The solution is that the electrons are in a **stationary state**. They form a kind of cloud that spins (has angular momentum) but is so symmetrical that it does not change shape as it spins, i.e., the charge distribution is the same at every moment. But for this it is necessary that the electrons do not behave like point-like particles, but like extended clouds.

In addition it became clear that these clouds occur only in very specific forms and states. Not any arbitrary angular momentum was allowed, not any arbitrary average distance from the atomic nucleus, not any arbitrary energy, but only specific discrete values. This could finally be explained by the fact that the electrons have certain oscillation properties, something wave-like, and these oscillations have very specific frequencies in the cloud around the atomic nucleus.

Mathematically, the whole thing is described by the **Schrödinger equation** (set up by Erwin Schrödinger in 1925), the basic equation of QM. In this equation, an electron is represented by a **wave function**, a mathematical function that assigns a complex number to each point in space. What kind of physical quantity this complex number describes (i.e. what it "means") remains open at first. Actually, physical quantities are always described by *real* numbers, the appearance of *complex* numbers is already the first oddity here. The Schrödinger equation determines how this function changes with time. In very specific cases, the change over time consists of a pure "phase rotation" (a version of oscillations that is only possible with complex numbers), without the function changing its form. These are the steady states we are looking for in the atom. The Schrödinger equation also shows how each stationary state can be assigned an energy, and that energy is proportional to the frequency of the phase rotation.

As already mentioned, the whole chemistry can be derived from this logic; every chemical reaction is a proof for the correctness of Schrödinger's equation. This is actually the least problematic part of QM; everything to be said in the following does not affect chemistry very much.

Wave-Particle Duality

The problem is that electrons often behave like point-like particles and not like waves. If one determines the location of an electron by a measurement, then it is always only a single location, not an extended cloud or a wave. Can the electron then be both an extended wave-like object and a point-like particle at the same time? Even stranger, the position of an electron found by measuring its location cannot be predicted. Instead, one finds some probability distribution for its location (such a distribution can be determined experimentally by running the same experiment many times and doing statistics with the results). The probability of finding an electron at a particular location is proportional to the square of the magnitude of its wave function at that location. So the complex number that the wave function attributes to a location encodes, among other things, the **probability of** the electron being at that location.

So there are situations where it is crucial that the electron is distributed uniformly over a volume of space, for example in the atom where it forms a cloud around the atomic nucleus. In this case, the wave function tells us, first, the density of the cloud at each location, and hence how the electric charge of the electron is distributed in the atom. Secondly, it describes an oscillatory behavior. If the electron is not bound to an atom, but whirls around freely, then this oscillatory behavior can lead to interference phenomena, as is known from other waves in optics or acoustics. In such situations, the **wave character of** the electron comes to the fore. On the other hand, there are situations in which the electron appears as a point-like particle **(particle character),** and in this case the wave function is the measure of the particle's probability of being present at a given location. The electron is a **quantum object,** it combines in itself both wave and particle properties, sometimes revealing one rather than the other. This is the so-called **wave-particle duality** in QM. The wave function, interestingly, is responsible for both: In one case it describes the wave itself, in the other case it provides probabilities.

There are experiments that combine wave and particle character. These are particularly taxing on common sense. The best known of these experiments is the **double slit,** in which an electron is shot through two narrow slits in a wall. On the other side, at some distance from the wall, is a screen that registers the point of impact of the electron. On its way, the electron behaves like a wave; it passes through both slits at the same time, spreads out on the other side, and interferes with itself. When it hits the screen, it behaves like a particle; it arrives at exactly one point (even though it was just distributed like a wave). If you repeat the experiment many times, you find that the statistical distribution of the points of impact is just the square of the magnitude of the wave function that results from the electron interfering with itself in a wave-like manner. This experiment is a classic. You can find much more detailed presentations and explanations in the literature or on the Internet.

In the above variant of the experiment, the electron as a wave passes through both slits at the same time. What happens if one sets up an electron detector at each of the two slits? Do the detectors respond to both, because the electron takes both paths at the same time? No, because detectors represent a location measurement in the area of the slits, so the electron already behaves like a particle at this point, and only one of the detectors reports its arrival. Which of the two detectors this is depends on chance. The wave function has the same amplitude at both slits, therefore the probability is exactly the same at both.

It is even possible to decide only at the last moment, when the electron is already on its way, whether the detectors are to be moved quickly in front of the slits or whether it is to be left with the screen at some distance, as in the first variant of the experiment. In this case, the electron on its way does not yet "know" whether it has to target both slits at the same time in a wave-like manner or only one of them in a particle-like manner (a so-called **delayed-choice experiment**). Nevertheless, at the moment of measurement, it will behave as expected.

In a similar manner, even more sophisticated experiments can be designed, which are aimed at tricking the electron (or photon, as we will see in a moment). And one would like to trick it, because this sudden switching from wave to particle seems to us to be absurd. If you want to know more about these experiments, do some research on the Mach-Zehnder interferometer or read one of Anton Zeilinger's excellent books.

Long before the Schrödinger equation was established, Max Planck (1900) and Albert Einstein (1905) found that light (or electromagnetic radiation in general) only occurs in[3]"quantized" form. The energy of light is composed of certain portions ("quanta"), and the energy of a single portion is proportional to the frequency, as is the case with electrons. The only difference is that the frequency of the electron is the frequency of something very abstract, namely the wave function, whereas the wave character of light is already evident from *classical* electromagnetism, Maxwell's theory: According to this classical theory, light consists of wave-like propagating vibrations of the electric and magnetic field. So it is the frequencies of these vibrations which determine the size of the energy portions.

With the help of suitable detectors, position measurements can be carried out on the individual energy portions. It turns out that the portions behave in exactly the same way as an electron: as soon as the location is measured, the portion, which just a moment ago behaved like an extended wave, immediately shrinks to a point, and the function which just described the wave-like distribution of electric and magnetic field now describes the probability to *which* point the portion shrinks. Light thus also consists of quantum objects, the **photons,** which have both wave and particle character, analogous to the electron; here, too, the wave becomes the measure of the probability of presence. The double-slit experiment can be performed with photons in the same way as with electrons.

This has a certain irony, because it was a centuries-long dispute whether light consists of waves or particles, until Maxwell with his theory, which

---

[3] Einstein got the Nobel Prize for his work on this subject, not for SR or GR.

united all the findings on the subject in great splendor, decided the dispute in favor of the waves. At last everything was clear, the great controversy settled once and for all: light consists of waves, not particles. The opponents gave up. And now it is all undone and turned into a compromise: Light consists of both waves and particles; it depends on the kind of measurement you make of it.

Wave-particle duality is universal, everything consists of quantum objects; the difference between matter, force fields and radiation disappears. Photons and electrons are just examples that are particularly important, but there are numerous others that we will come to. Unfortunately, a somewhat confusing inaccuracy has crept into common usage: People generally speak of elementary particles instead of elementary quanta. However, elementary particles are quantum objects, they have both wave and particle character.

The quantum nature of photon *and* electron also explains why certain substances absorb and emit light at very specific wavelengths. According to the Schrödinger equation, the stationary states of an electron in an atom have very specific energies ("energy levels"). To get to a higher level, the energy difference between the two states must be absorbed, for example in the form of a photon that has exactly that energy. And a photon with a certain energy has a certain frequency and therefore a certain wavelength. If, on the other hand, the electron falls to a lower level, a photon with the corresponding energy (namely the energy difference between the two levels) is emitted.

Chance

With QM, chance also seems to have entered physics at a fundamental level. All the theories we have discussed so far have been deterministic. From the exact state of the world at a single point in time (positions and velocities of all classical particles; electromagnetic field at every point in space) follow with mathematical precision the exact states of the world at every other point in time. The universe runs like clockwork, there are no chance and no freedom.

In QM, however, chance seems to be real (we leave freedom out of it for now). At which point of space a concrete electron or photon appears is in *principle* subject to chance, the statistical probabilities resulting from the wave function. This property of QM met with great suspicion, especially from Einstein. For him, one certainty was *"God does not play dice!"*, one of his most famous sentences. Therefore he was convinced until the end that QM could not be the last word.

Over time, three fundamentally different attitudes emerged towards chance in QM:

1. Chance is real and cannot be attributed to any ignorance on our part. God does play dice. This is the position of the **Copenhagen interpretation,** the view of a majority among the pioneers of QM.
2. Chance is only an illusion, which, as in classical physics, comes from the fact that we do not know the exact state. In reality, there are **hidden variables** (also called hidden parameters) that must be added to QM, making it a deterministic theory again. If these mysterious variables were not hidden from us, we could predict with certainty at what exact point the electron or photon would appear. But over the decades, as the consequences of QM were thought through more and more – a lengthy process, since QM is so contrary to our learned habits of thought that even mathematically very simple conclusions were simply not seen for decades – this possibility seemed less and less likely. It became apparent that hidden variables would have to have some very messy and implausible properties in order not to conflict with the tested predictions of QM. The only variant of QM of this kind that can still boast a certain following today is so-called **Bohmian mechanics,** but we will not discuss it here.
3. Chance is only an illusion, which however is based on a completely different cause than in the previous point. In reality, the electron (or photon or whatever) is still distributed over the space, even if the position is measured. It only appears to us as if the result of the spatial measurement is a single point. This is the approach of the **many-worlds interpretation,** which we will discuss in detail below.

Measurement Problem and Uncertainty Principle

Another feature of QM that is widely regarded as unattractive is the **mystification of the measurement process.** From classical physics and common sense, one is used to the idea that when a certain property of a certain object is measured, that property has already been there. The measurement just brings it to light. When I read the temperature of my living room on a thermometer, I don't assume that my measurement creates the temperature, it just shows it to me. The room has the temperature regardless of whether there is a thermometer there or not. Furthermore, I know that the measurement itself follows known physical rules; so it is also clear *how* the room temperature causes my thermometer to display the appropriate numerical value.

This is no longer true when measuring the location of a quantum object. The point-like location, indicated by a click or flash of a detector, is a property that the object did not have at all before the measurement, because before that the object (e.g. electron, photon) was extended and wavelike. It seems that the detector *creates* the property of the object having a particular location in the first place. How can this be? It is also completely unclear *how* this happens physically. How can the detector contract the extended object into a point? This contraction is also known as the **collapse of the wave function.** So what exactly happens in this collapse, when and how does it take place, and when and how does it actually decide at *which* point it takes place? This series of questions forms the famous **measurement problem of** QM.

This also includes the fact that the experimenter seems to decide with his measurement *which* properties he helps to exist in the quantum object. There are, in fact, properties that are mutually exclusive. When we measure the location of a quantum object, we ensure that it has a well-defined position in space. We bring out its particle character. To do this, we have to accept that the object has a completely "fuzzy" momentum, and in fact the more precisely we measure the location, the fuzzier the momentum becomes. We recall that the momentum of an object is defined as mass times velocity, at least for non-relativistic objects, i.e. objects whose velocity is much less than the speed of light. If we restrict ourselves for the moment to electrons (photons always travel at the speed of light), then we can also interpret the fuzziness of momentum as the fuzziness of velocity. The mass of an electron, on the other hand, is uniquely determined in every case. If we measure the momentum instead, we bring out the wave properties of the object. This then blurs the location. The electron becomes an electron cloud. The more precisely we measure the momentum, the more blurred the location becomes and thus the more extended the electron cloud.

The location and momentum of a quantum object are mutually exclusive, they are **complementary** – a term coined by Niels Bohr for this: Both properties are in principle inherent in the quantum object, but the experimenter must decide which of the two to bring to light, thereby excluding the other. The extent to which the precision of the two properties are mutually exclusive is determined mathematically in **Heisenberg's uncertainty principle.** Besides location and momentum, there are other pairs of complementary properties that are mutually exclusive.

Originally, many thought that the uncertainty principle came about through the interaction of the measuring device with the quantum object, through a kind of recoil. That is, it was thought that the quantum object had both properties, but the interaction required to measure one property changed the other in such an uncoordinated way that it was no longer possible to make a statement about it. However, this is a misinterpretation. It can be shown that the assumption that both properties of a complementary pair are given simultaneously leads to contradictions, completely independent of the action of a measuring device.

State Space and Schrödinger's Cat

From the wave function results the probability distribution of the measured location in the case of a location measurement, but also the probability distribution of the measured momentum in the case of a momentum measurement can be derived from it, as well as the probability distributions for many other properties potentially to be measured. Other properties, such as mass and electric charge, are always uniquely determined. However, there are other properties, such as the *spin of* an electron (a property we can think of in a somewhat vague analogy as "self-rotation"), that also exhibit fuzziness and therefore must be specified in terms of probabilities, but which cannot be derived from the wave function. In order to accommodate these properties in QM as well, the concept of the wave function must be extended to the even more abstract **state vector.** The state vector is an element of a certain mathematical structure, the so-called **state space of** the quantum object. The state vector contains the wave function as one component, but it also contains other components that describe, for example, the spin of the quantum object (or its probability distribution). In the literature, wave function and state vector are often not separated. Often one says "wave function", where actually a state vector is meant.

The Englishman Paul Dirac introduced a notation for state vectors in the 1930s which is still used today, namely $|\cdots\rangle$, where for $\cdots$ any symbols or labels can be used which characterize the state. Let us illustrate this notation with the help of **Schrödinger's cat.** To sum up the madness of QM, Erwin Schrödinger invented the thought experiment of a quantum mechanical cat, which has achieved a high fame even among non-physicists. This cruel thought experiment involves a cat being locked in a box where it will be killed by a quantum mechanical process coupled to some killing mechanism with a 50% probability. Since this is supposed to be a quantum mechanical cat, its state is described by a state vector containing probability distributions for all potential properties of the cat in its current state. Since we are concerned here exclusively with the property "dead or alive", we denote by $|\,dead\rangle$ the state vector of a dead cat, and by $|\,alive\rangle$ that of a living one.

A crucial property of vectors is that they can be added. So we can mathematically form the state |dead⟩ + | alive⟩. This state vector describes the superposition of a dead and a living cat. That |dead⟩ and |alive⟩have equal proportions of this sum implies, by the rules of QM, that a measurement of the "dead or alive" property will produce a dead cat with 50% probability and a live cat with 50% probability. This is exactly the superposition condition described in the thought experiment. A "measurement" in that case might be for the experimenter to open the box and see if the cat is dead or alive. Now the amazing thing about QM is that it is the measurement that "brings out" this property. Before the experimenter opens the box, the cat is, depending on the point of view, simultaneously dead and alive, or neither dead nor alive.

Much of the confusion caused by such a description comes from the fact that **two different languages** are needed to describe a quantum mechanical process. First, there is the language used to represent the experiment. This handles objects that we can see, touch, and locate in space, in this case a box, a cat, and the technical objects used to potentially kill her. Apart from a few technical terms, this language is by and large identical to our everyday language. Second, there is the purely mathematical language of state vectors that "live" in an abstract state space. The connection between the two languages is given solely by the fact that state vectors can be used to predict probabilities for the outcome of the experiment.

We have already discussed this "bilingualism" in detail in Chap. 6, namely for the *whole of* physics, not only for QM. In fact, it already exists in classical physics. Experimental facts have to be translated into the mathematical language of theory, and vice versa. The difference between classical physics and QM, however, is that in classical physics the translation causes much less problems. It leaves our conception of the world intact. The mathematical language of classical mechanics is about masses distributed in space, in the same space that constitutes the world in our imagination (in our imagination, the world is space with the totality of "things" in it). Each piece of mass thereby has a set of clearly defined properties that are uniquely given at any point in time. This is how we imagine the world. This gives rise to the illusion that the mathematical objects with which the theory deals completely represent the things of the world one to one and are in their totality quasi equivalent to the world.

In QM, however, this illusion is broken. The mathematical objects of the theory "live" in a completely abstract state space, which is quite different from the space that constitutes the world in our imagination. The translation between the mathematical objects of the theory and the experimental facts or "things" of the world is complicated. The connection underlying this translation is ultimately only that the mathematical objects can be used to calculate probabilities for outcomes of experiments. But whereas in classical mechanics we could still say with reasonably good conscience that the things around us "consist of" the mass points of the theory, in QM we can no longer say that things "consist of" state vectors. The two languages remain separate.

This separation is only one of the ways in which QM does violence to our intuition. Another is that it tears a crack in our conception of material things. It shows that at certain times or in certain constellations things do not have the unique properties that we intuitively attribute to them. This is the case with our quantum mechanical cat, which is in a superposition state of dead and alive. However, it is only a thought experiment. For a real cat such a state is impossible. Indeed, it turns out that such superpositions only hold if the object is extremely well isolated from its environment, much better than you could ever isolate a cat. You would have to surround it with vacuum and cool it down to absolute zero temperature, and as you can easily imagine, even then it would not be $|dead\rangle + |alive\rangle$, but only remain $|dead\rangle$. The need for such strict isolation also poses great challenges to the development of quantum computers. However, in a thought experiment, one can overlook this difficulty to illustrate a point of the theory.

A "measurement" of the cat's state, i.e. when the box is opened, then determines whether the cat is dead or alive. This moment also represents the transition between the two languages: The mundane (or "classical") view of a cat with certain properties is not accurate before this moment. Instead, its state must be represented in an abstract state space. After this moment, however, the mundane view is appropriate. The cat is either dead or alive. This transition in quantum mechanical experiments is also known as the **Heisenberg cut.** But when exactly does it take place? At the moment when the lid of the box lifts? The moment the image of a cat, living or dead, appears on the experimenter's retina? At the moment when the information enters his consciousness? This question can be asked for all quantum mechanical experiments: When exactly is a measurement considered to have occurred? It is one of the puzzles of the measurement problem. In order to approach it, we have to turn to another great curiosity of QM: entanglement.

Entanglement

An electron (or its spatial distribution) is described by a wave function. You probably think that with two electrons there are of course two wave functions. But this is wrong! While for a single electron there is one wave function in a three-dimensional space, for two electrons there is still only one wave function, but in a six-dimensional space. For three electrons, it is still only one wave function, but in nine dimensions. Why is that? Because the probabilities of presence of the electrons are not independent of each other. If there were two wavefunctions, each would describe the probability of presence of the respective electron, *independently*. But electrons are negatively charged and repel each other. Therefore, the probability of finding the second electron at a location is much lower if the first is just near that location. So we have to calculate with the *combined* probability that electron 1 is at position A *and* electron 2 is at position B, and this can only be done by accommodating both positions (six coordinates, namely three for each of the two electrons) in a single wave function.

This mutual dependence is called **entanglement.** The oddities arising from this aspect of QM are even greater than those mentioned so far, such as the uncertainty principle and wave-particle duality. For ultimately all particles are somehow connected to all others via chains of interactions. A solid body is solid because all its atoms are held "in line" in a long chain of electrical forces. If we push the body at its left end, not only does its left end move in the direction of the push, but through the chain of interactions the push is passed through to the other end, the body moves as a whole. In liquids and gases, particles are constantly colliding, influencing each other. Even our perception only works because we are connected to things through interactions. The light they emit or reflect, for example, hits our eyes and produces a reaction there.

It follows that everything is ultimately entangled with everything else, so that, strictly speaking, all particles must be brought together in a single wave function, the **wave function of the universe.** For this, it is not even necessary that every particle interacts directly with every other particle. It is sufficient if particle 1 interacts with particle 2, particle 2 with particle 3, particle 3 with particle 4 and so on. Already this causes that no group of particles is completely independent from the rest of the world and thus a total wave function of the universe has to be used.

Why then do the calculations with wave functions, which refer to single electrons, often work so well nevertheless? In fact, they are always only approximations. Individual particles can – at least for a certain period of time – be isolated from the rest of the world much better than a macroscopic object, and for this period of time they are described very accurately by separate wave functions. As soon as an interaction takes place, however, we have to resort to a common wave function of all particles involved in the interaction.

If two particles interact with each other at a certain time, it can happen that their entanglement is maintained for a long time afterwards. This is partly due to the conservation laws that are still valid in QM, for example the conservation of momentum: When two particles interact with each other, the sum of their momenta is conserved. In so-called *scattering experiments,* two particles are shot at each other, they interact with each other in a certain range and time period, after which they fly apart in different undetermined directions determined only by probabilities. By the law of conservation of momentum, the directions the particles take after colliding depend on each other, they are entangled. For example, if one particle flies up to the left, the other must fly down to the right. But QM now requires that the direction of each of the two flights remains indeterminate until it is measured.

Suppose a ring of detectors is built around the experiment. Before the particles reach the detector ring, they are described by a common wave function. This contains a part in which the first particle flies to the upper left and the second to the lower right, but also a part in which the first flies to the upper right and the second to the lower left, and many other parts. Only at the moment when the particles reach the detector ring is it decided which path they have actually taken. Because of the entanglement, the decisions of the two particles are dependent on each other. If the detector at the top left reports the arrival of the first particle, the detector at the bottom right must report the arrival of the second. If, on the other hand, the detector at the top right reacts, this must also apply to the detector at the bottom left.

The crux is that QM requires that the direction of the particles is not known until immediately before they hit the detector ring. Because of the dependence by entanglement, the decision must fall on both sides at the same time. This stretches our imagination to a great extent: How can the lower particle "know" for which direction (left or right) the upper one has decided? In the meantime, the particles are a good distance apart and no longer have any possibility of exchanging information. The decision seems to be made "globally", on both sides at the same time, coordinated with each other, but paradoxically without such coordination actually taking place. It was Albert Einstein who first pointed out this behavior; he spoke of a **spooky action at a distance.** To this day, more and more sophisticated experiments have been set up demonstrating this spooky action at a distance in ever more impressive ways. It is also a crucial component of new technologies such as quantum cryptography and quantum computing.

Copenhagen Interpretation

Entanglement also sheds a whole new light on the quantum mechanical measurement process and the problems associated with it. However, let us first summarize the view presented so far, which is known as the **Copenhagen interpretation of** QM. The name comes from the fact that it essentially dates back to the collaboration between Niels Bohr and Werner Heisenberg in 1927, while Heisenberg was a guest of Bohr at his institute in Copenhagen. In the now famous *Bohr-Einstein debate*, Bohr subsequently defended this interpretation of QM against Einstein, who found it implausible and contradictory, which he supported among other things by his example with the spooky action at a distance. In fact, the term "Copenhagen interpretation" is not very well defined. Several variants of it circulate, differing in their focus and some details. It is used to summarize a whole range of views from the time of the pioneers of QM. I want to present here, what is the core of this interpretation from my point of view, but I have to point out, that some interpret the interpretation a little bit different (the interpretation of the Copenhagen interpretation is itself a wide field).

In my view, the core of the Copenhagen interpretation is that science is an activity of man and must also be understood as such. Man does not know the true reality underlying his theories and his own actions. He develops theories based on experiments and uses the theories to predict results of further experiments. In QM this prediction succeeds only in the form of probabilities, one can make only statistical statements. It turns out that in QM, to a much greater extent than in other areas of physics, two different languages are needed to describe theory and experiments. The language of experiments is essentially the language of classical physics and is about material objects in space and time that are put together to form an experimental setup, where an object takes on the role of a measuring device that essentially measures a property of the object under study that can be expressed in the language of classical physics, such as the location or momentum of an electron or the polarization of a photon. The language of theory, however, is about wave functions or state vectors that live in a completely abstract state space. Neither language represents absolute reality. In particular, wave functions or state vectors are only tools that give the physicist the best possible prediction. The collapse of the wavefunction at the moment of measurement is not something that "actually" happens (though there are variants of the Copenhagen interpretation that see it differently), but only represents the boundary where we need to update this tool with the new information gained during the measurement.

The separate language of theory is necessary because some properties of the quantum objects under investigation are *complementary* – such as location and momentum – i.e., both properties are potentially inherent in the object, but only one of them can be determined at any given time – a circumstance that cannot be correctly expressed in the language of classical physics. In this complementarity, the observer (or experimenter) has a crucial role to play. It is up to him to decide which of the two complementary properties he wants to bring out in the object, which one he wants to help to concretize. His experimental setup, the measurement process, which is initialized by this, determines the character of the object (e.g. wave or particle), which thereby changes from a possibility to something given.

Many-Worlds Interpretation

This view appears too mystical to most physicists today. In fact, what actually happens during the measurement process remains completely open. One could get the idea to say: Well, maybe it is just beyond human possibilities to understand the measurement process itself, we have to accept the modesty of the Copenhagen interpretation. But this is wrong. The measurement process can be understood and demystified in a very concrete and dynamic way by describing the measuring device and observer themselves in terms of quantum mechanical state vectors. This is where the concept of entanglement helps us.

To illustrate this, let us take the example of a **qubit,** a quantum object with a binary property that can take only two different values when measured. Qubits can be realized in many ways. A well-known example is the spin of the electron, which, measured along a certain axis, has only two possible expressions ("spin-up", "spin-down"). Let us call the state vectors representing these two values $|Q0\rangle$ and $|Q1\rangle$. The value of this binary property is to be measured with a measuring device. For this we now need to introduce state vectors as well. Let us say, $|M-\rangle$ is the state vector of the device before the measurement, $|M0\rangle$ the state vector of the device when it displays the value 0, $|M1\rangle$ the state vector when it displays the value 1.

Let us assume that before the measurement the qubit is in the state $|Q0\rangle$. Then the total state of qubit and measuring device is given by $|Q0\rangle\,|M-\rangle$. During the measurement, the qubit interacts with the measuring device (otherwise the measuring device could not measure anything). The dynamics of this interaction can be fully described in the language of state vectors, using the **Schrödinger equation**, which is responsible for the time evolution of all state vectors. At its end, i.e., after the measurement, there is the state

$|Q0\rangle |M0\rangle$, i.e., the interaction has caused the device to display the value 0, matching the qubit that is actually in the state $|Q0\rangle$. If, on the other hand, the qubit is in the state $|Q1\rangle$, then the initial state is $|Q1\rangle |M-\rangle$, and the dynamics according to Schrödinger's equation leads to a final state $|Q1\rangle |M1\rangle$ after the measurement, the measuring device displays the value 1.

The game can be continued with the observer (experimenter) if we introduce state vectors for him as well. Let's say $|B-\rangle$ is the state of the observer before the measurement, $|B0\rangle$ the state of the observer when he has read the value 0 from the device, $|B1\rangle$ the state when he has read the value 1. (The quantum mechanist often simply says "state" instead of "state vector", but must keep in mind that by "state" he means precisely this abstract mathematical construct). The reading of the device also takes place through an interaction, this time between the device and the observer (light is emitted from the device and strikes the observer's eye). This can now also be fully described in the language of state vectors, using the Schrödinger equation. For a qubit in state $|Q0\rangle$, the sequence for the total state consisting of qubit, measuring device and observer is now as follows:

1. $|Q0\rangle |M-\rangle |B-\rangle$: At the beginning, neither the measuring device nor the observer knows the state of the qubit.
2. $|Q0\rangle |M0\rangle |B-\rangle$: After the interaction between qubit and measuring device, the measuring device knows the state of the qubit. However, the observer has not yet read the device and is therefore still in the initial state
3. $|Q0\rangle |M0\rangle |B0\rangle$: After the interaction between the measuring device and the observer, i.e. the reading, the observer also knows the state of the qubit.

The interactions thus involve a transfer of information. Let us assume that outside the laboratory a colleague of the observer is waiting for the observer to tell him the result of the measurement. Then we could also describe this colleague by a state vector, completely analogous to the observer himself, and the communication of the result as an interaction between observer and colleague, described in the language of state vectors, by the Schrödinger equation. Isn't this a wonderful explanation of everything? Doesn't this take the whole "mysticism" of the Copenhagen interpretation *ad absurdum*? Doesn't it also make the strict separation of the two languages superfluous by simply treating *everything* in the language of state vectors?

It's not quite that simple. The weird thing happens with superposition states. Suppose the qubit is in the state $|Q0\rangle + |Q1\rangle$, that is, a superposition of 0 and 1. Then the initial state is $(|Q0\rangle + |Q1\rangle) |M-\rangle |B-\rangle$, which is the same as $|Q0\rangle |M-\rangle |B-\rangle + |Q1\rangle |M-\rangle |B-\rangle$. Now the Schrödinger equation has the property that the constituents of a sum of state vectors evolve completely independently. That is, the constituent $|Q0\rangle |M-\rangle |B-\rangle$ evolves just as we discussed above, namely via the intermediate state $|Q0\rangle |M0\rangle |B-\rangle$ on to $|Q0\rangle |M0\rangle |B0\rangle$. Completely analogously, the second component of the sum, $|Q1\rangle |M-\rangle |B-\rangle$, evolves via the intermediate state $|Q1\rangle |M1\rangle |B-\rangle$ to $|Q1\rangle |M1\rangle |B1\rangle$. The final state is therefore $|Q0\rangle |M0\rangle |B0\rangle + |Q1\rangle |M1\rangle |B1\rangle$.

What does this mean? The qubit, the measuring device and the observer are entangled with each other, their states are no longer independent of each other but related to each other. The zero part of the qubit includes a measuring device that displays the value 0 and an observer who has read the value 0. The one part of the qubit includes a measuring device that displays the value 1 and an observer who has read the value 1. The overall state after the measurement is a superposition of these two options. Originally, the superposition was in the qubit alone; its state had a zero and a one component. Through the interaction, first between the qubit and the measuring device, then between the measuring device and the observer, the information (0 or 1) of the respective portion is passed to the measuring device and then to the observer. However, this also transmits the superposition state. The measuring device is now "split" into a measuring device that indicates 0 and one that indicates 1. The observer is split into one that reads the value 0 and one that reads the value 1.

So something quite different happens than in the Copenhagen interpretation. There, each measurement had a single result, an unambiguous result. Now the measurement has *both results at the same time*. The observer himself has become a split personality, with his two parts (the one that read 0 and the one that read 1) henceforth evolving completely independently of each other, without noticing anything about each other. The two variants of the observer will in turn interact with their environment, communicate with it. Thus the splitting is carried further and further outwards, so that finally the whole world is covered by it. The whole world is then a superposition of a world in which the value 0 and one in which the value 1 was measured.

In the case of Schrödinger's cat, it is the same: the moment the observer opens the box, he splits into an observer who sees a dead cat and one who sees a living cat. Through further interactions of the observer with his environment, the whole world splits into one in which the cat is dead and one in which it is alive.

Most measurements allow for more than two possible outcomes. In that case, the world splits into as many variants as the measurement has possible outcomes. You might think that in many cases this is an infinite number, because there is often a continuous spectrum of possibilities, for example when we measure the location of a particle. But in fact every measurement has a finite resolution, you never get an arbitrarily precise result. This finite resolution then determines how many different branches the world branches into. For example, if you determine a position along a ten centimeter axis with a resolution of a tenth of a millimeter, there are 1000.

But does that really make sense? When we make a measurement, we actually see only one result, and not several at the same time. Are we so schizophrenic that we simply don't *realize* that we see all possible results at the same time? But the Schrödinger equation predicts exactly that: the different branches of the world's events are henceforth completely independent of each other. Each part of the observer therefore feels itself to be the only one, it doesn't know anything about the existence of the others.

But since we cannot *prove* that all the other variants of the world or of ourselves are also there, it is an *interpretation of* QM that this is so, and not a scientific statement. It is the so-called **many-worlds interpretation.** It was first formulated in 1957 by Hugh Everett in his dissertation. The work met with complete rejection at the time. It was not until the 1970s, inspired by further work by Bryce DeWitt (who also coined the name *many-worlds interpretation*), that it gained wider recognition and has since enjoyed increasing popularity.

I want to emphasize here that the many-worlds interpretation is the simplest and – up to a point – most plausible interpretation of QM:

- Since measuring instruments and observers are composed of atoms, i.e. quantum objects, it is quite correct to regard them themselves as quantum objects, which are therefore correctly described by state vectors.
- The Schrödinger equation has proven to be the correct equation describing the time evolution of quantum objects in all areas of QM.
- Applied to measuring devices, observers and their environment, the Schrödinger equation unambiguously predicts a splitting of the global state vector; the many-worlds interpretation is thus a direct consequence of the Schrödinger equation.
- The many-worlds interpretation describes ("explains") the entire measurement process dynamically with the help of a single equation. It thus describes *how* the measurement physically takes place. In the Copenhagen

interpretation, on the other hand, the measurement process remains completely mysterious.

- There are also many other physical processes in which entanglement takes place quite similarly to a measurement process. So the measuring process is additionally demystified by the fact that it is a physical process like many others.
- The forced "bilingualism" of the Copenhagen interpretation is omitted. Everything is accurately described in the language of state vectors.
- The *spooky action at a distance* is also disenchanted by the many-worlds interpretation. A coordination of entangled objects over large distances is no longer necessary. Let us assume that the quantum objects A and B are entangled with each other, so that a certain measurement result X at A always occurs in combination with a measurement result Y at B. If A and B are far away from each other, this was called a spooky action at a distance: it looked as if the two quantum objects had to communicate over the distance so that the right combination always occurred. But in the many-worlds interpretation, all possible measurement results actually take place. In this case, the values that belong together via entanglement land in the same branch of the world state, so they are always observed together. For landing in the same branch no remote tuning is necessary, it results simply from the dynamics of the Schrödinger equation.
- Chance, which was such a thorn in Einstein's side, thus disappears from QM. Quantum mechanical events are *perceived* as random because we simply do not notice that all other possible events have also taken place, just in a different branch of the world state.

So up to a certain point, the many-worlds interpretation is the simplest and most plausible interpretation of QM. But it too has its limitations. A frequently voiced criticism is that the probabilities, which play a central role in the Copenhagen interpretation and which we can actually reproduce experimentally, cannot be derived from the many-worlds interpretation. To reproduce a probability experimentally is to do statistics and look at relative frequencies. Suppose we have prepared a qubit in a certain way, and QM (in the Copenhagen interpretation) predicts that given that initial state, the probability of measuring the value 1 is 75%. Then we can check this prediction by running the same experiment many times. If we run it 100 times, and get the value 1 73 times, we will consider this prediction confirmed (we don't expect to get *exactly* 75 hits, because there is always some statistical fluctuation). If, on the other hand, we get the value 1 only 50 times, we will consider the prediction to be disproved. There are clearly defined procedures in statistics to decide when a statistical statement is to be considered confirmed and when it is to be considered disconfirmed, within certain limits (it is very unlikely, but not impossible, to get only 50 hits out of 100 trials with a probability of 75%).

In the view of the many-worlds interpretation, the state of the world splits into two separate branches on each trial, one for each of the two possible outcomes. On each trial, the number of branches doubles; if we start with just one branch, after 100 trials we have 2^{100} branches, a huge number with about 30 decimal places. Now the problem is that in these 2^{100} branches, on average, 50 times the value 0 is obtained and 50 times the value 1 is obtained. In almost all branches, therefore, observers will conclude that the probability of the value 1 is 50%, not about 75% as predicted by the Copenhagen interpretation and as we find in actual experiments.

Chance, in the many-worlds interpretation, is a priori something subjective. In reality, all possibilities actually take place. But each part of the observer *perceives* the possibility realized for him as the only one, and as one that came about by chance. The probability he assigns to this accidental result is also subjective. In order to correctly reflect all of the *actual* experimental ratios found in QM, the many-worlds interpretation must make a *prediction about* what probabilities observers subjectively assign to outcomes. This prediction must be consistent with the probabilities of the Copenhagen interpretation, because it is these probabilities that we actually find in results after many experiments. But now it turns out that the many-worlds interpretation predicts that in the vast majority of branches, observers arrive at different results (namely, 50% instead of 75% in our example). There *are* such branches in which the number of ones found is compatible with the 75% probability, but they are strongly in the minority. The more experiments one conducts, the more they are in the minority. From the point of view of the critics of the many-worlds interpretation, this shows that it arrives at false results and is therefore not really an interpretation of QM at all, but represents a different theory (since it makes different predictions), and one that has been disproved by the experimental facts.

The problem with this line of reasoning is that it is not at all clear how we are to imagine the wandering of our subjective sense of self through the many-worlds branches. After a measurement, I find myself, from my subjective point of view, in one of the branches, each representing a particular outcome. I am (or feel I am) "the same me" as before, and I know nothing of other me's that have ended up in the other branches, although they too exist according to the many-worlds interpretation. Critics of the many-worlds interpretation now say that it is a problem that only a numerically small fraction of the branches are compatible with the probabilities found experimentally. But is this really a problem? It would be one if all branches were completely equal prior to our subjectively finding ourselves in one of them. Then we would subjectively move through the branching tree of possibilities completely at random, and typically end up in such branches (simply because they are heavily outnumbered) in which we conclude the probability of the value 1 is 50% rather than, say, 75%. But it's not at all clear whether the branches are equal in terms of our subjectively finding ourselves in them, or how to judge that in the first place. So the argument with probabilities hangs a bit in the air from my point of view.

Max Tegmark has proposed an experiment by which everyone can convince himself of the correctness of the many-worlds interpretation. This experiment plays with finding oneself subjectively in one of the branches. The trick is to apply Schrödinger's cat experiment to yourself, over and over again, that is, to play Russian quantum roulette. Expose yourself to a device that kills you by some quantum process with a 50% probability. According to the Copenhagen interpretation, with a 50% probability, you're done for, and with a 50% probability, you're lucky and live on (assuming you perceive that as luck). According to the many-worlds interpretation, after the experiment you are both dead and alive, in two separate branches of world events. However, in the branch where you are dead, the subjective self no longer exists for you to find yourself. Therefore, by necessity, you find yourself in the branch in which you are alive.

Perform the experiment many times in a row. According to the Copenhagen interpretation, you are sure to be dead after a few tries (that's just the way it is with Russian roulette). However, if the many-worlds interpretation is correct, you will find yourself alive every time. It will seem like a miracle to you, but it is just the logic of the branching tree of possibilities that have all become real. Try it out, take a chance! For some truths, you have to face death.

Well, maybe you will reconsider, because I will now explain to you why the many-worlds interpretation is wrong. For us physicists, dealing with QM in practice is characterized by the fact that we permanently jump back and forth between the two languages ("classical" language of "things" in space and language of state vectors in state space). In doing so, the state vectors have a meaning for us in that they refer to the "things" of the classical language; they express statistical statements about their properties. We express these references by giving the state vectors certain labels. For example, we use the label $|Q1\rangle |M1\rangle |B1\rangle$ for a state vector to refer to a qubit, a measuring apparatus, and an observer, and the 1 symbolizes in each case a particular property that we can observe about these things. (The qubit is not, strictly speaking, a "thing" of classical language, but the 1-property we measure on it has its meaning precisely because it can be measured in a certain way, by the measuring apparatus, that is, by a "thing" in space).

The many-worlds interpretation now says that one gets to the bottom of QM by using *only* the language of state vectors. But with this one deletes all references which give the state vector its meaning in the first place. The state vector no longer describes the state of anything, it is a completely independent object. If we now designate a state vector with $|M1\rangle$, then it is no longer because it refers to a measuring device with a certain property, but because its intrinsic behaviour and its interaction with the observer's state vector creates in the observer the *impression of* a thing called a measuring device. But also this sentence is not correct, because also the state vector "of the observer" does no longer refer to a "thing" (or even a "subject") called observer, but only creates the *impression of* representing an observer by its intrinsic behaviour or by its interaction with further state vectors.

It is even more complicated: Due to the global entanglement of everything with everything, there is actually not even a separate state vector $|M1\rangle$, but strictly speaking only the state vector of the entire universe, let us call it $|X\rangle$. Strictly speaking, in the many-worlds interpretation one can only deal with the dynamics of this global state vector. The entire "classical world of things" must *emerge in* a most complicated way from the behavior of $|X\rangle$. But perhaps this is possible?

No, it's not. The problem can be broken down into two parts:

1. Without any recourse to the "classical" language, how is one to find a meaningful decomposition of the global state space $Z_{Universe}$ into "smaller" state spaces,

$$Z_{Universe} = Z_{Object1} \times Z_{Object2} \times \cdots, \tag{7.3}$$

whose elements can then be interpreted afterwards as states of individual objects, as "particles" or other objects in a "world"? Only such a decomposition allows us to interpret the global state vector as composed of *constituents*, e.g. as

$$|X\rangle = |Q0\rangle |M0\rangle |B0\rangle |R\rangle, \tag{7.4}$$

where R here stands for "rest of the world".

2. Once such a meaningful decomposition has been found, any temporal development of $|X\rangle$ can be interpreted as an interaction between the individual constituents. Now the question arises how these interactions lead to the fact that certain constituents appear to us ("us" now also means certain constituents of the global state vector) as objects with certain "classically" describable properties.

This second question can indeed be answered: Once the state space has been decomposed into constituent parts (factors), then from their interactions (which are described, as always, by the Schrödinger equation) follows a mechanism called **decoherence**, the description of which goes beyond the scope of this book, but which leads us to see – in each branch of the constantly splitting many-worlds - a particular object with particular properties.

However, the really hard part of the problem is the first one, the decomposition of the state space. I argued in Schwindt (2012) that it is insoluble. The global state vector in the global state space simply does not contain enough information to give us a clue as to how to perform this decomposition. It seems completely arbitrary. But it is only after this act of arbitrariness that the global state vector tells a story of "things" in a "world". So the content of this story, the predictions about the behavior of "things", do not come from the behavior of the global state vector – as the many-worlds interpretation would require – but from a structure arbitrarily superimposed on it, a decomposition of the global state space. The many-worlds interpretation can therefore not deliver what it promises.

If you have read Duhem, it seems methodologically questionable from the outset. Duhem emphasizes that in physical theories there must always be a translation between the mathematical terms of the theory and the observations of the experiment. The Copenhagen interpretation clearly defines how this translation has to take place, i.e. how the state vectors relate to certain facts of an experiment. In this sense, physics is always *bilingual;* this bilingualism is only particularly prominent in QM due to its special properties. The many-worlds interpretation now tries to resolve this bilingualism by translating the observer of an experiment and the whole process of observing itself into the abstract language of theory. In doing so, however, it ultimately produces only a mathematical construct that *no longer means anything,* from which no physical meaning can be squeezed out even through calculations.

Too bad. At the beginning we still said that the many-worlds interpretation is *up to a certain point* the simplest and most plausible of all interpretations of QM. Perhaps it has some value after all? Indeed, it can be taken as a reasonable *complement to* the Copenhagen interpretation. In the Copenhagen interpretation, the question was at what point in a measurement process the *Heisenberg cut* occurs, i.e., up to what moment the language of state vectors is to be applied (in physicists' jargon: when the "collapse of the wave function"occurs). The many-worlds interpretation provides a perspective from which to say: It is arbitrary. The collapse of the wave function is not something that actually *happens,* it is just an expression for our transition from one language to another. But *when exactly* we transition from one language to the other doesn't matter. The measuring apparatus and the observer can also be expressed by state vectors, and the measuring process itself can be understood as a quantum mechanical process. It's just that at *some point* we have to make the transition from one language to the other, we have to *relate* the state vector to something we observe, because by itself it *means nothing.*

But what is "real" now? The many worlds of the many-worlds-interpretation are an idea, which the mathematics of QM evokes in us. But the classical *one* "world of things" is also only an imagination that our everyday mind forms. How exactly these imaginations relate to reality we do not know. But can we perhaps say a little more about it? For example, can we find out if either of the two conceptions (many worlds vs. one world) is closer to reality? Yes, run the Russian quantum roulette! If you find yourself still alive after 1000 tries, you can say with certainty that there is something to the many worlds. (At ten tries you could still refer to your personal guardian angel, but 1000 tries, even the most persistent guardian angel won't do it). Is there a less life-threatening way to find out? I don't know.

7.7 Quantum Field Theory

QM, as we discussed it in the previous section, is initially *non-relativistic,* i.e., it does not yet take into account SR and certainly not GR. Wave functions describe the distribution of a quantum object in *space;* the evolution of a state with *time* is described by the Schrödinger equation, whether or not the state is expressed in terms of a wave function. Space and time seem to have little to do with each other here. Moreover, due to the *non-local* character of QM, expressed for example in the **spooky action at a distance**, the limit of the speed of light seems to be violated: For entangled states of two quantum objects, the decision which of the possible states is realized takes place simultaneously on both sides, no matter how far apart they are spatially. In the context of relativity, the word "simultaneous" should immediately ring alarm bells: Absolute simultaneity is not defined there at all.

The latter problem is solved, in my view, by the many-worlds interpretation, which describes how the decision for a state in the form of entanglement with observers comes about *dynamically* over a period of time, and precisely not instantaneously. As discussed, in my view the many-worlds interpretation is not a truth in itself, but it has a special perspective to offer, which makes a contribution to the understanding of QM, and with this contribution in particular the problem of spooky action at a distance is solved (Sect. 7.6).

The former problem, namely the non-relativistic character of the Schrödinger equation, was already addressed in the late 1920s by extending this equation so that it exhibited the typical relativistic spacetime symmetry. But this gave rise to new problems, inconsistencies, which eventually showed that the whole notion of what a particle (or quantum object) is had to be modified one more time. The result was relativistic QFT.

Already in classical, i.e. Newtonian physics, we had made the transition from classical mechanics, populated by isolated point-like particles, to classical field theory, in which the entire space is occupied by fields. A similar transition is now taking place in quantum physics. QM deals with individual quantum objects, each of which replaces the classical particles, in the sense of *reduction by replacement*. These quantum objects (which, as I said, are still called particles due to a linguistic blunder) have certain states that describe probability distributions for certain properties that we can measure in experiments. Many of these properties have a direct correspondence in the world of classical particles, for example location and momentum.

QFT deals with **quantum fields** which *replace* the classical fields, in the sense of *reduction by replacement*. The same formal rules apply to these quantum fields as in QM: The quantum fields are in certain states, which are elements of a certain state space. Such a state determines probability distributions for certain properties of the quantum field, which can be measured in experiments. The problem, however, is that the state space of QFT is much more difficult to understand than the state space of QM. That is why problems in QFT are rarely treated on the basis of the general state space, but still on the basis of particles, only that the word "particle" now means something else again.

In QM, a particle is a quantum object that is in a certain state. In QFT, a particle is itself a state, namely a certain state of a quantum field. The quantum object is now the quantum field, and the fact that there is a particle characterizes a certain state of that quantum field. Thus, one *object* has become the *state of* another object.

To make this a little clearer, let's think about the states of an electron in an atom. As you already know, an electron in the shell of an atom can only have very specific energies. Let's say it's in the state with the lowest energy. (In old-fashioned language, that would mean it's in the innermost shell). Now it can be *excited* out of this state into a state of higher energy, via the intake of energy, for example, by collisions or by absorption of light. From this higher state it can again be excited by a similar process into an even higher state. However, it can also fall back to the lower state by releasing energy.

In quantum fields, the state with the lowest energy is called **vacuum.** Through various kinds of interactions, the quantum field can be excited out of this state into a state of higher energy. This state is then called "a particle". A further excitation transfers this into an even higher state with the name "two particles" and so on. Conversely, a two-particle state can "fall back" into a one-particle state by releasing energy, as can a one-particle state into a vacuum. In QFT, therefore, the number of particles is constantly changing.

Since there are superposition states in QFT as well as in QM, states with different numbers of particles can be superposed. A state of the quantum field can be a bit of a vacuum, a bit of a one-particle state and a bit of a two-particle state. Now the number of particles is a quantum property like any other and can usually only be given in terms of probabilities.

The wave function of a particle, from which probability distributions for location and momentum can be derived, further specify a state. That is, there are infinitely many possible one-particle states (the vacuum state can be excited to a one-particle state in an infinite number of different ways). What we know from QM as a wave function specifies in QFT *which* one-particle state we are talking about.

For photons, this view is relatively obvious. We already know about the "existence" of an electromagnetic field and are familiar with the fact that a photon is a kind of quantum embodiment of this field. Now what about the vacuum here? In everyday language we use the word "vacuum" for a volume of space in which there are no particles of matter (atoms or their constituents: electrons, protons, neutrons). With this meaning there can be a ray of light in a vacuum, that's why we often talk about the speed of light in vacuum. In the sense of QFT, however, the vacuum state of the electromagnetic quantum field is given by the fact that all values of the field are equal to zero everywhere with a probability of 100%, because then the energy of the field is at its lowest. In this sense, "vacuum" also means the absence of light. If the electromagnetic field is now excited by interaction with an electric charge, this happens precisely because the charge "emits" a photon. If many photons are emitted in the same direction by such reactions, they form electromagnetic radiation, i.e. what we perceive as a beam of light, provided that the wavelength (which is fixed in the concrete photon state) is in the visible range of the spectrum.

For electrons, however, the view of QFT is surprising. For it follows that there must be a so-called electron field whose excitation is the electron. Unlike the electromagnetic field, the electron field does not appear in the form of a classical, macroscopic field, but in the form of small groups of electrons, which are usually bound to atomic nuclei.

An important prediction of QFT is the existence of **antimatter.** For every charged field there are two different single-particle excitations, one with positive and one with negative charge, but otherwise with identical properties, especially with identical mass. To the negatively charged electron there must therefore exist a positively charged twin brother, an **antiparticle,** the so-called **positron,** which is also an excitation of the electron field. This was already predicted by Dirac in 1929 and demonstrated a little later; a first great success in this early phase of the development of QFT.

Both QM and QFT are quantum theories, i.e. they use the quantum formalism with its states and probability statements and the associated difficulties of understanding as far as the measurement process is concerned. QM deals with particles that *have* states, QFT with particles that *are* states, namely states of fields. QM can thereby be reduced to QFT, in the sense of *reduction by replacement.* There are various types of fields that QFT can be about, and these can be fully classified within the framework of QFT. Thus QFT is still a fairly general formalism. Only very specific fields occur in nature, and by specifying them, one concretizes the general formalism of QFT to a particular manifestation. The specification that deals with the electron field (and thus its particle states, the electrons and positrons), the electromagnetic field (or its particle states, the photons) and their interactions is **quantum electrodynamics (QED).**

The simplest QFT that can be formulated about the electron field and the electromagnetic field, is not QED, but a theory without interactions. In this minimal theory, the two fields coexist peacefully without affecting each other. A world governed by this theory would be extremely boring. It would be a world without electromagnetic forces. The state of the fields would never change. A constant number of photons, electrons and positrons from the beginning to the end of time would pass through space with eternally constant momenta, without paying attention to each other and without anything ever happening.

In QED, on the other hand, there is an interaction between the fields that takes the form of violent reactions at the level of particle states: Electrons and positrons can excite the electromagnetic field, thus producing photons that can then be reabsorbed by another electron or positron. The momentum of the electrons and positrons changes in a way that, from a classical point of view, looks as if a Newtonian force had acted, emerging from a classical electromagnetic field. In this way, classical electromagnetism is reduced to QED (again, in the sense of *reduction by replacement*).

It becomes particularly explosive when an electron meets a positron: the two annihilate each other, producing a photon whose energy is just equal to the sum of the energies (including mass) of the electron and positron (the law of conservation of energy also applies here). So in this process, the number of particles that were derived from the electron field has decreased by two, but the number of photons has increased by one. Because only one electron and one positron are annihilated at a time, the difference between the number of electrons and the number of positrons remains constant. (This is only true within the framework of this theory. If we add in other interactions that go beyond QED and involve other fields, the picture changes somewhat, but even then it turns out that the difference between the number of electrons and positrons can only have fluctuated slightly over almost the entire history of the universe). Today, we find huge numbers of electrons throughout space, but virtually no positrons. This can be interpreted to mean that there must have been an excess of electrons in the early universe. Electrons and positrons annihilated each other at that time, and only the excess of electrons remained. The same is true for other particle-antiparticle pairs, such as protons and antiprotons. If there had been a surplus of positrons and antiprotons left over instead, we would be made of them today, and all the names would be reversed: the electrons and protons would then be antimatter for us.

Since QED is time-reversal invariant, to every process also the opposite process is possible. So if an electron and a positron can annihilate each other producing a photon, then it must also be possible for a photon to decay into an electron and a positron (because that's exactly what it looks like if you run the electron-positron annihilation backwards). The only condition for this, because of conservation of energy, is that the energy of the photon must be at least twice the mass of the electron. This is several orders of magnitude higher than the energy of visible light photons. Today such energetic photons are rare, but in the hot early days of the universe they were ubiquitous, constantly creating new particle-antiparticle pairs. Only when the universe had cooled down to the point where such energetic photons were severely thinned out could a permanent electron-positron annihilation begin.

QED can be generalized from the electron to all other electrically charged particles, in particular also to the proton. However, protons are more complicated, firstly because they are not elementary particles (they have a substructure of so-called quarks), and secondly because they are mostly bound in atomic nuclei, for which another interaction, namely the strong nuclear force, is responsible.

The Mathematical Problems of QFT

QFT is very demanding from a mathematical point of view. Many aspects of it are still not understood. There are actually only two types of situation in which the handling of it is somewhat clarified: scattering processes and decays. One speaks of a scattering **process** when two (in some cases more) particles move towards each other, react with each other when they meet (collide), and then the particles that emerge from this encounter fly apart again. The particles that emerge from the encounter may be the same particles as before the encounter, e.g. two electrons simply deflecting each other from their original paths due to electrical repulsion. Or new particles may emerge from it because quantum fields are excited to a higher number of particles by the reaction. The new particles can arise in addition to the old ones or replace them.

In a **decay,** you have one particle at the beginning, which then disappears at a certain time and is replaced by several new particles (one quantum field falls back to a lower energy level, others are excited to a higher level). As usual, the time of the decay can only be predicted in terms of probabilities. Usually it is given in terms of a **half-life**, which is the time at which the particle has decayed with a probability of 50%. You know this from radioactive decays of atomic nuclei. Here we have the same thing, only on the level of elementary particles.

Both types of processes, scattering processes and decays, have one thing in common: the interaction is active only for a very short period of time, namely at the moment of collision or decay. Before and after that, the particles of the initial or final state are either alone or far away from each other without interacting.

What we are much less comfortable with in the context of QFT, however, are bound states such as atoms, where cohesion comes from forces acting *permanently*, not just for a brief instant. The simplest atom, the hydrogen atom, which consists of only one proton and one electron, is one of the easiest exercises in QM. All physics students learn how to calculate its states in their basic course on QM, as an example of what QM can do. In QFT, however, the hydrogen atom is a decidedly difficult case. We know that QM can be reduced back to QFT, in terms of *reduction by replacement*. But to do this reduction in the case of the hydrogen atom is a surprisingly awkward challenge that still awaits an elegant solution.

For scattering and decay processes there is a very successful method, the so-called **perturbation calculus,** to predict the result. "Result" here means: a probability distribution for the particles present after the process, as well as their energies and momenta. The method consists of assuming a certain initial and final state and summing up the various intermediate states via which the initial state can reach the final state, from which it can then be concluded at the end with what probability the initial state reaches the chosen final state.

In 1949, Richard Feynman developed a scheme, which has become famous, to visualize the calculations resulting from this method: the **Feynman diagrams.** A scattering process is represented by an infinite number of diagrams. Each diagram represents a particular mathematical expression that describes a particular path from the initial state through a particular set of intermediate states to the final state. Thus, each diagram represents one of the infinite number of expressions to be summed up in the perturbation calculus. An example of such a diagram is shown in Fig. 7.8.

We are dealing with four different levels here:

1. At the foundation there is the actual QFT, a theory that deals with quantum fields and their states.
2. Building on this, there is an approximation method, the perturbation calculus, which can be used to approximate probabilities for scattering and decay processes. It is an approximation method because we would have to do an infinite amount of computation to calculate all of the infinite number of expressions relevant to a single scattering process. One must therefore restrict oneself to a few expressions that are assumed to be the main contributors. This works surprisingly well, which is quite a miracle, as we will see in a moment. The results of these calculations can then be compared with experimental results.

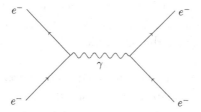

Fig. 7.8 Simple example of a Feynman diagram. Two electrons (e⁻) are scattered by each other. The diagram symbolizes a certain mathematical contribution to the scattering, which can be understood as the exchange of a single "virtual photon" (γ). The time runs from the bottom to the top

3. The mathematical expressions can be visualized by graphical symbols, which are assembled into Feynman diagrams. This visualization helps the physicist purely psychologically in calculating and communicating. The diagrams serve as a reminder of which expressions one has to work with at the moment, and can be conveniently drawn during lectures, thereby "illustrating" the mathematical process one is describing.

4. Finally, there are the metaphors from everyday language that we use both among physicists to communicate with each other and in conversation with lay people to whom we try to explain the content of our work. These metaphors are often borrowed from the objects of classical physics. For example, we speak of "particles" being created or annihilated. This is a metaphor or, if you like, a "nickname" given in QFT to a certain variety of mathematical expressions. By using this nickname, a certain association is evoked in the listener or reader, a certain image that forms his imagination from the term "particle". This image, however, usually has little to do with what the mathematical expression actually says.

The particles of classical mechanics are reduced to the quantum objects of QM, then even further to certain states of quantum fields. Since in each case it is a *reduction by replacement*, the character of the objects changes completely in the process. Nevertheless, the term is passed down from the classical level to QFT and continued to be used as if it were not something quite different. We will come back to this in Sect. 9.2.

A justification of this use of language arises solely from the functioning of the reduction: the states of the quantum fields behave in certain situations in such a way that they simulate the behavior of a classical particle.

The perturbation calculus has two major problems: infinity and infinity. Namely, one quickly finds that the calculus yields infinity for almost every summand (that is, for almost every Feynman diagram). There are then some tricks, small modifications to the calculus, that can be used to eliminate these infinities and instead give each diagram a concrete, finite numerical value. At first, it was not clear why these tricks would actually work. Only in the course of the following decades did the connections behind them become clearer and clearer.

The formalism that is used here is called **renormalization.** It makes use of the fact that every experiment, every measurement, has a limited spatial resolution and, in this sense, takes place on a certain **scale** given by the degree of resolution. An optical microscope, for example, works with visible light, electromagnetic radiation of a wavelength around 500 nanometers. This light reflects off the structure you want to examine in a certain way, creating the image we see in the microscope. The light microscope has a natural limit: the wavelength of the light. Structures smaller than about 500 nanometers cannot be resolved by the light microscope. Therefore, the *scale of* the light microscope is said to be 500 nanometers. The electron microscope, which uses electron radiation with a wavelength of about 1 nanometer, can resolve structures down to a minimum size of 1 nanometer; this is the *scale of* the electron microscope. The formalism of renormalization indicates that the *scale of* a process must always be taken into account and that the parameters of a theory also depend on the *scale*. These parameters include, in particular, the "constants" that determine the strength of an interaction. So in QFT these "constants" are no longer constant at all, but depend on the scale.

There is a certain kinship between the subject of renormalization and the mathematics of *fractal geometry*: if you ask how long the coast of Norway is, the answer must be: That depends on the *scale* at which you are measuring, that is, the spatial resolution you want to consider. A coastline is a *fractal* structure: the closer you look, the more bays and fjords become visible, each increasing the length of the coastline by its "squiggle". As you keep resolving, zooming deeper and deeper into the structures, more and more sub-structures will become visible, little twists and turns of the coastline, indentations within indentations, and so on, and the answer to the question of the total length of the coastline will keep increasing. It is similar with the structures of QFT, and the formalism of renormalization shows in which way. A meaningful answer is always obtained only if one assumes a finite resolution. The size of this resolution is usually fixed by the experimental conditions. In this way the infinities in the individual summands of the perturbation calculation are eliminated.

Once you have eliminated the infinities by renormalization, however, you are still left with a problem: The individual summands are now finite, but their total sum is still infinite. A sum with infinitely many summands is called a **series** in mathematics. A series can yield a finite value (it is said to **converge**) despite having an infinite number of summands, for example $\frac{1}{2} + \frac{1}{4} + \frac{1}{8} + \frac{1}{16} + \cdots = 1$. Here, each summand is only half the size of the previous one, and the consequence is that all the infinitely many summands together give only the value 1. The crucial thing here is that the summands get smaller and smaller fast enough. The summands of the perturbation calculation do indeed become smaller and smaller at first, the sum seems to *converge* towards a certain value. Unfortunately, it can be shown (as Freeman Dyson already found out in 1952) that they also become larger again after a certain point and that the series does not converge in the end.

Surprisingly, physicists were only slightly disturbed by this realization. At that time, renormalization was not well understood either, they only knew the computational tricks to get finite values, but without understanding why they actually worked. It was then found that using these tricks, and considering only the first few summands of the series (the first few Feynman diagrams), one surprisingly obtained correct predictions for scattering and decay processes. The predictions were not only correct, but even incredibly precise, accurate to many many decimal places, the most accurate predictions ever made in the natural sciences.

You have to let this roll off your tongue: The most precise predictions of all time were accomplished with (from the perspective of the time) unmathematical witchcraft, with forbidden tricks that seem to work for some inexplicable reason. In the metaphysical imponderables of QM we had identified an origin of the "shut up and calculate" mentality. In the sorceries of perturbation calculus we find the second, perhaps more formative, origin. The calculation yields the experimentally reproduced result, so better not think too much about it.

Fortunately, QFT did not stop at this level. With renormalization the solution of the first problem with infinity was understood. The second problem, the lack of convergence of the perturbation calculus, even then attracted surprisingly little attention. It was hardly mentioned in introductory courses on QFT. It was rather mathematicians who understood the problem and provided answers why the whole thing works anyway. For mathematicians and physicists, I recommend the instructive yet entertaining paper *How I Learned to Stop Worrying and Love QFT* (Helling, 2011). His summary begins with the words:

The material in these notes will not be useful for any concrete calculation in quantum field theory that a physicist might be interested in. But they might give him or her some confidence that the calculation envisaged has a chance to be meaningful.

An important step beyond "Shut up and calculate".

In summary, the perturbation calculus method works, gives extremely accurate predictions, and there are good mathematical reasons *why* it works on closer inspection (on superficial inspection however it looks like witchcraft).

The formalism of renormalization describes how the details of a theory change as one changes the *scale* (resolution) at which one views it. In the simplest case, this simply means that certain parameter values of an interaction change continuously as the *scale* is changed continuously. But there are also more dramatic changes, called **phase transitions,** where the entire behavior of the fields changes abruptly.

Moreover, renormalization provides transitions from fundamental to **effective theories,** where the quantum fields of a more fundamental theory are replaced by quantum fields of a less fundamental theory. For example, we know that protons and neutrons are in some sense composed of even "smaller" particles (again, in a very metaphorical way of speaking), the **quarks.** These are described by a QFT in which the quark particles are excitations of quark quantum fields.

However, unlike electrons or photons, quarks can never appear individually, but only in combination, for example in the form of protons and neutrons. If the scale of an experiment is larger than the size of the proton or neutron[4], then there is no way to detect the substructure from quarks. In this case, one can perform calculations based on a QFT that deals with proton and neutron quantum fields, not quark quantum fields. This theory gives accurate predictions as long as the scale of the processes studied is larger than the proton and neutron, so that, relative to the given resolving power, the proton and neutron appear to be point-like. It forms an *effective theory,* a QFT, which is not fundamental, but only valid as an approximation, which is valid in the context of certain scales. Conversely, it may happen that a QFT that we still consider fundamental today ultimately turns out to be an effective theory that no longer holds below a certain length scale, where it must be replaced by another, more fundamental QFT. It may even happen that below a certain length scale we encounter a theory that is no longer a QFT at all.

QFT and GR

In contrast to pure QM, QFT includes SR. A big open question is whether it is also consistent with GR, which is the theory of gravity and curved spacetime. This issue can be looked at in two steps. First, one can restrict oneself to doing QFT on a curved *background.* That is, one takes into account that spacetime is curved, but does not yet include any influence on the curvature by the quantum fields, as actually envisaged by GR. This first step works well, but once again leads to new surprising effects. These cause the concept of a "particle" to move *even* further away from what we intuitively imagine it to be. Now it can even depend on the reference frame whether the state of a

---

[4] We mean by larger scale a larger *length scale*; in quantum theories each length scale has associated with it a momentum scale; a higher value for the length scale means a lower value for the momentum scale and vice versa. Therefore, when one speaks of "larger" or "smaller", one must always add whether one means the length scale or the momentum scale. The energy of a particle is related to the momentum, which is why we often speak of an energy scale. Here we always mean the length scale, because it is easiest to understand.

quantum field is recognized as a state with particles or as a vacuum. This effect becomes relevant when an observer is subjected to extremely high accelerations (the so-called **Unruh effect**), but much higher than could be achieved by humans, as well as in the vicinity of black holes, where the spacetime curvature is very high. The "evaporation" of black holes, also known as **Hawking radiation,** is based on this.

If one then wants to include the influence of the quantum fields on the spacetime curvature in the second step, it becomes even more complicated. Then one has to raise the metric (the mathematical object from which the curvature is calculated) itself to a quantum field. For a quantum-theoretic superposition of different states of quantum fields must lead to a quantum-theoretic superposition of different spacetime curvatures, and this can only be accomplished if the spacetime geometry itself becomes a quantum object. If one tries this, one quickly finds that the method of perturbation calculus applied to this new quantum field does not work.

In their minds, many physicists equate QFT and perturbation calculus quasi with each other. That is why it was sometimes wrongly concluded from the non-functioning of the perturbation calculus that **quantum gravity** could no longer be a QFT. But this is not correct. A QFT of gravity is still conceivable, just not with the method of perturbation calculus. There are different approaches to this, but all of them are not quite unproblematic so far. In addition to the conceptual and mathematical difficulties, there is the fact that the "natural" scale from a mathematical point of view, on which the effects of such a theory typically become noticeable, is the **Planck scale,** which is about 10^{-35} meters, trillions of times smaller than we are able to resolve with the best instruments. In other words, we cannot get any experimental clues as to how the theory we are looking for behaves, and if we have constructed a candidate theory from theoretical considerations, we cannot test it experimentally. The conditions for a solution of this fundamental open problem of physics therefore look decidedly unfavorable. We will come back to this in Chap. 8.

7.8   The Standard Model of Particle Physics

The matter of which we are made and which surrounds us consists of electrons, protons and neutrons, whereby protons and neutrons together form the atomic nuclei, and electrons form the atomic shells. The fourth particle we encounter in everyday life is the photon, the "building block" of light. In addition to these four types of particles, however, there are numerous others. Why do we notice almost nothing of these? There are four different reasons:

1. Most particles are unstable and decay very quickly. Already the neutron is unstable if it is "free", i.e. not bound in an atomic nucleus. A free neutron decays with a half-life of about 15 min into three other particles, namely into a proton, an electron and a neutrino. Almost all other particles found in particle physics decay within tiny fractions of a second, much too fast to catch them in the wild or do anything with them.

2. Some particles hardly interact with us. This is the case for the neutrino, for example. Every second, several hundred trillion neutrinos pass through each and every one of us without us noticing. The interaction is simply too weak. It takes huge detectors to get a few neutrinos over very long periods of time to give the detector a signal of their existence. We know about the huge number of neutrinos more for theoretical reasons. We know that the sun has to radiate about 10^{38} of them every second to keep its energy balance. The presumed particles of dark matter (Sect. 7.9) also seem to affect us only via gravity. In this case, too, a large quantity of these particles could flow through us without us noticing anything.

3. Other particles are extremely rare. This applies, for example, to the anti-matter partners of electron and proton: They are stable and would react very strongly with us (with a big flash and mutual annihilation), if they were there. But they aren't there. Antimatter is extremely rare, it doesn't occur in space, and on Earth we can only make it by expending a lot of energy. It is assumed that in the early universe there was about equal amounts of matter and antimatter. But then, through collisions, they almost completely annihilated each other, leaving only a certain surplus of matter, and it is from this surplus that all that remains today is made up. For all we know, the extinction should have been even more complete. Much more matter remained than would be statistically expected. Why the excess was so great is one of the unsolved mysteries of cosmology.

4. Certain particles cannot be isolated. This is the case with the mysterious *quarks*. Protons and neutrons have a substructure; in a sense, they can be said to "consist of" smaller particles, namely three quarks. But this "consist of" is only a partially accurate description. For one thing, the proton and neutron are both about 100 times heavier than the three quarks combined. On the other hand, there is also no way to separate the quarks from each other, i.e. to split the proton (or neutron) into them. Both are due to the special properties of the strong nuclear force, by which the quarks are held together. Unfortunately, we cannot discuss this further here.

The term elementary particle is not entirely straightforward. For example, let's look at the definition on the German Wikipedia page (as of June 2019):

"Elementary particles are indivisible subatomic particles and the smallest known building blocks of matter." This is a very problematic definition. First of all, you have to specify what you mean when you say a particle is "small". An electron in the atomic shell behaves like an extended cloud whose volume is huge compared to the atomic nucleus, although the latter is by no means elementary. An electron that is not subjected to any external forces can, in principle, grow to any size. Only when a *position measurement is* made does an electron shrink to a tiny volume, the size of which is determined by the accuracy of the measurement. An atomic nucleus, however, always retains a certain amount of expansion, even with an arbitrarily accurate position measurement, simply because it is composed of several more elementary particles that cannot simply be pushed on top of each other. So the only meaningful use of the word "small" is defined on the state in the case of a position measurement: Elementary particles are "point-like" according to an infinitely precise position measurement.

The use of the word "indivisible" in the definition above is even more questionable. A proton is not elementary, it has a substructure of quarks, but it is still indivisible because there is no way to split it into its constituent parts. Conversely, we know that most elementary particles decay in a short time; for example, there is an elementary particle called a muon that decays into an electron and two neutrinos in a few microseconds. This decay causes the muon to "split" in a sense, and its energy is completely divided among the three decay products. According to the definition above, the muon could not be elementary. The crucial point, however, is that the muon does not have any *substructure* of electrons and neutrinos *before* the decay, the muon is "point-like" in the sense just discussed.

Last but not least, "subatomic" is a misleading term. In what sense can we say that a photon is subatomic? "Subatomic" suggests that we are dealing with a component of an atom, or with something that is "smaller" than an atom. But most elementary particles have nothing to do with atoms, and we have already criticized the concept of smallness above.

Let's look at the definition from the English Wikipedia page for comparison (as of June 2019): *"In particle physics, an elementary particle or fundamental particle is a subatomic particle with no substructure, thus not composed of other particles."* Again, the term "subatomic" bothers us, but otherwise the definition is much better. It is really about the absence of substructures that appear when one spatially "scans" a quantum object, for example by scattering experiments.

The Particle Zoo

The only elementary particle that was already known at the end of the nineteenth century was the **electron.** The internal structure of atoms was unravelled more and more in the first three decades of the twentieth century. Around 1917, Rutherford found the **protons to** be positively charged components of the atomic nucleus. The missing mass had to occur in electrically neutral constituents, the **neutrons, which** were finally detected by Chadwick in 1932.

The history of the **photon** is complicated. It was postulated as a particle of light already by physicists of the seventeenth century, but of course without knowing anything about electromagnetism and QM. As a quantum object it was reinvented by Planck and Einstein at the beginning of the twentieth century.

The **positron,** i.e. the antimatter particle to the electron, was postulated by Dirac in 1929, in the context of his work on relativistic QM, a precursor to QFT. At about the same time, the first indications of a positively charged twin of the electron were found in experiments. The positron was definitely detected in 1932, the same year as the neutron.

In radioactive decays, which were becoming better understood in the late 1920s within the framework of QM, the conservation laws for energy, momentum and angular momentum seemed to be violated. After the reaction, something of all three quantities seemed to have been lost. Therefore, in 1930, Pauli postulated another very light electrically neutral particle, the **neutrino,** which was created during the reactions and immediately dissipated, balancing the conservation quantities. Because these particles interact so weakly with matter, they are extremely difficult to detect directly. But in 1956, this was finally achieved, and since then neutrinos have been part of the confirmed inventory of particle physics.

The heavier, unstable particles are typically created in violent collisions of other particles and decay again immediately afterwards. During their short existence, however, they can be made to leave characteristic traces using suitable experimental methods. Further conclusions about the properties of a short-lived particle can be drawn from the decay products detected by detectors. The only problem is that the collisions needed to produce such particles must sometimes be *very* violent. There are two known ways to get at such collisions:

On the one hand, there is **cosmic radiation** (not to be confused with cosmic *background radiation,* which is still to be discussed; Sect. 7.9). These are charged particles (protons, electrons, atomic nuclei) which come from space with high energy and hit the Earth, where – still in the atmosphere – they collide with terrestrial atoms and produce cascades of other particles which fly away from the site of the collision like sparks, among them also the exotic short-lived particles which otherwise do not occur in nature.

On the other hand, humans can do it themselves, with the help of **particle accelerators**. Here, charged particles such as electrons or protons are accelerated with electric fields to high energies and focused by magnetic fields on certain paths, where they are brought to head-on collision with other particles, again with the result that cascades of other particles are created in the process.

Particle accelerators have been constantly developed over the last 90 years, producing particles with ever higher energies, up to the (to date) last masterpiece, the Large Hadron Collider **(LHC)** at the CERN research institute near Geneva. As a result, in the early years of particle physics, cosmic rays were the main source of new discoveries, but from the 1950s onwards they were increasingly replaced in this role by particle accelerators.

The first new particle found as a collision product of cosmic rays was the **muon** in 1936, a kind of big brother of the electron, i.e. it has essentially the same properties as the electron, except that it is heavier and unstable. It was followed in 1947 by the pion and the kaon – particles with completely different properties, which at first could not be properly classified.

In the 1950s and 1960s, the new particle accelerators were a success: more and more new particles were discovered, and there was talk of a veritable **particle zoo.** The zoology of particles has – similar to zoology in biology – nested group terms analogous to classes, orders, families etc.. Thus most of the particles discovered from 1950 onwards belonged to the class of **hadrons,** relatives of the proton and the neutron and, as it turned out, of the kaon and the pion. The diversity of these hadrons was so great, and at the same time they were so closely related and could be grouped so systematically, that it was reasonable to hypothesize that they consisted of a small number of even smaller particles, named **quarks** by Murray Gell-Mann. The behavior of these quarks is very complicated and quite different from that of the other known particles. It is described to a large extent by a particularly complicated QFT: **quantum chromodynamics (QCD)**. It represents the so-called **strong nuclear force.**

But quarks also interact via the **electromagnetic force** and via the **weak nuclear force.** Electrons and muons interact with the rest of the world via electromagnetism and via the weak nuclear force, but not via the strong nuclear force. Neutrinos interact exclusively via the weak nuclear force.

Actually, the concept of force at this level has little in common with the Newtonian concept of "mass times acceleration". A force here is rather a general reaction scheme, according to which particles act on each other or are transformed into other particles. So this concept too has undergone a transformation as it descended through the hierarchy of theories.

Every force is described at this level as being transmitted by a carrier particle. The electromagnetic force is transmitted by photons, the strong nuclear force by so-called **gluons.** Photons and gluons are massless and therefore travel at the speed of light. The weak nuclear force also has carrier particles, the so-called **W** and **Z bosons.** However, these have mass and therefore move slower than the speed of light. They also decay very quickly into other particles. Therefore, the weak nuclear force has a very short range. The direct detection of W and Z only succeeded in 1983, but the indirect evidence for their existence was already so strong 10 years earlier that hardly anyone doubted them.

It turned out that electromagnetism and the weak nuclear force can be covered by a common theory, the **electroweak theory.** At low energies, the two forces have very different properties. But the electroweak theory now states that the properties converge when the particles involved in the interaction have very high energies. The two different forces become a single force. Similar to QCD, the electroweak theory also ensured that many strange relationships in the particle zoo suddenly made sense.

However, the theory has a catch: it is only consistent if all particles that belong to the class of so-called **fermions** and are subject to the weak interaction are massless. This includes electrons, muons and quarks, which were already known to have mass after all. So was the electroweak theory wrong? Everything else fitted together far too well for that. However, a way out of this dilemma was found: the mass of electrons etc. is not mass at all, but only looks like it. In reality, the apparent mass is only an **interaction energy** between the electron (or muon, etc.) on the one hand and a newly postulated quantum field, the **Higgs field,** on the other – more precisely with its vacuum state. Since this vacuum has always and everywhere the same properties, also the interaction energy and thus the apparent mass of the electrons (e.g. muons, quarks) has always and everywhere the same value.

The Higgs field, like any other quantum field, has excited states, namely the **Higgs particles.** In the late 1960s, when this theory was established, it was all still mere theory. Experimentally, there was already a lot to be said for the electroweak theory; the only weakness was the existing mass of the fermions. The Higgs field was a speculative solution to this problem, but there was no evidence for it. It was not until 2012 that the Higgs particle and thus also the existence of the Higgs field were proven at the LHC, so the theory found its glorious confirmation.

Electroweak theory and QCD together form the so-called **Standard Model of particle physics.** The origin of both theories lies in the 1960s, but already in the mid-1970s the evidence was so good that large parts of the community of particle physicists regarded it as a general consensus. Since then, the Standard Model has become an unprecedented success story. The missing evidence was added (in particular W, Z and Higgs particles as well as an additional quark species), experiments with ever higher energies and ever greater precision confirmed it to more and more decimal places.

On the one hand, this is a great triumph; on the other hand, it leaves particle physicists somewhat perplexed as to what to do next. The intensity with which theory and experiment constantly cross-fertilised until the 1970s and led to new ideas, theories and discoveries was immense. Hopes were raised that the last riddles of the universe would soon be deciphered on the basis of particle physics. But since then, the pace has slowed considerably. New ideas did emerge, new theories for new, as yet undiscovered particles were proposed, but could not be confirmed in experiments. Everything kept falling back to the Standard Model. The remaining open questions of physics, however, are still unresolved. We will come back to this in Chap. 8.

Connection Between Standard Model and QFT

The standard model of particle physics is a QFT. All particles are excited states of quantum fields. QFT is the theoretical framework, the formalism, in which all particle physics operates. Strictly speaking, QFT represents a catalogue of possible variants of particle physics, and the Standard Model is a concrete manifestation of it. Theoretically, one can come up with an infinite number of conceivable particle physics. By saying that it should be a QFT (because all experiments indicate that this is the correct theoretical framework), one already severely restricts the choice. Only a few types of particles and interactions ("forces") remain.

To get from QFT in its most general form to a concrete manifestation like the Standard Model, one has to go through essentially the steps mentioned below. But first of all two remarks on this:

1. Each "determination" in the following steps is made by comparison with experimental results. The theorist is, of course, free to invent his own theory, carrying out the determinations as he sees fit instead, deviating from

those of the Standard Model. This too has value: the real world can be compared with *possible* worlds to see where we stand in this space of possibilities.

2. If you are not already a professional, it will be difficult for you to understand what the steps mean in detail; I can only give you a vague scheme here, a somewhat foggy impression.

So here are the steps that lead from general QFT to a particular manifestation like the Standard Model:

1. Determination of certain **charge schemes** (the technical term for this is *gauge groups*): A charge scheme largely determines the interactions present in the framework of the theory. The electromagnetic interaction, for example, involves electric charge, with which we are all more or less familiar. Electric charge is a one-dimensional scheme: we can specify charge in terms of a single number, a multiple of the elementary charge. For example, an electron has the charge -1, a positron +1, as does a proton. Neutrons and neutrinos are electrically neutral, so they have the electric charge 0. Now the remarkable thing about QFT is that by this charge scheme, the entire properties of the electromagnetic interaction are already fixed. There is only one way in which such charges can react with each other. It follows also automatically the existence and the properties of the photon which transmits this interaction.

   The strong nuclear force, on the other hand, follows a three-dimensional charge scheme: a charge with respect to the strong nuclear force must be given in terms of three values, which have been called (in a purely metaphorical way; there is no relation whatsoever with optical colours) red, green and blue charge. From this scheme already follow the whole properties of QCD, including those of the gluon which transmits this interaction.

2. Definition of **quantum fields:** In the case of the Standard Model, these are a) three so-called lepton fields, to which the electron field and the muon field belong, b) three neutrino fields, c) six quark fields, and d) the Higgs field. From the existence of the fields follow also the existence of the respective particles, e.g. the electron as excitation of the electron field, as well as the existence of the associated antiparticle, e.g. the positron.

3. **Assignment of quantum fields to charge schemes:** This is a mathematical procedure (in technical language: assignment of quantum fields to representations of gauge groups). In the end, it means that each particle is assigned all values to all charge schemes chosen in point 1, but not any arbitrary combination is allowed, but certain rules have to be considered.

It also follows from the assignment which particles are excluded from which interaction. If a particle, e.g. a neutrino, has no electric charge (i.e. the electric charge is 0), it is excluded from the electromagnetic interaction. If it has no red, green or blue charge with respect to the strong interaction, it is excluded from the strong nuclear force (this applies to all particles except quarks).

4. Determination of **masses:** In the Standard Model, only the Higgs particle has a mass, all other particles are massless.

5. Determining the **strength of the individual interactions:** These include the strength of the electroweak interaction and the weak nuclear force, but also the strength of the interactions between the fermion fields and the Higgs field (so-called *Yukawa couplings*). Since the strength of the interaction, as described in Sect. 7.7, depends on the scale, this determination must be carried out at a *certain* scale.

6. Definition of a scheme for the so-called **symmetry breaking:** In the example of the Standard Model, this scheme shows how the electroweak interaction becomes two separate interactions, namely the electromagnetic force and the weak nuclear force. In symmetry breaking, the Higgs field also acquires a special vacuum state, which leads to the aforementioned mechanism of fermions acquiring an apparent mass given by the strength of the Yukawa coupling defined in point 5. The stronger a particle interacts with the Higgs field via the Yukawa coupling, the heavier it appears.

With these steps, the general theoretical formalism of QFT is transferred into a concretely realized variant, namely the Standard Model. In the process, many things seem arbitrary. Why just these charge schemes? Why just these fields? Why just these values for the strength of the interactions? These are the open questions, with which the Standard Model leaves us and which we want to take up again in Chap. 8.

7.9 Cosmology

Cosmology is the science of the universe as a whole, its size, origin, evolution, and future destiny. Cosmology is an age-old part of philosophy; it is one of the things to which our philosophical wonder and questions are most directed. It is all the more gratifying that, with the resources of astronomy and physics, we are now able to provide very precise answers to some fundamental questions of cosmology. However, there are also many unresolved questions. Of the unresolved questions, there are some that, with luck and skill, can still be resolved, and others whose answers will probably remain hidden from us forever, either for practical or for fundamental reasons.

Roughly speaking, the findings of cosmology can be summarized in three sentences:

1. The universe is very big.
2. In fact, it's getting bigger.
3. In fact, it's getting bigger and bigger faster and faster.

Let's go through these findings in order.

The Universe Is Very Big

Since the Copernican revolution in the sixteenth century, man no longer believes himself to be at the centre of the universe. He knows that the Earth is only one of eight planets revolving around the Sun. Once there were nine, but Pluto had to give up its prominent status as a planet and, demoted to a dwarf planet, devotes itself only to its private life.

It also quickly became clear that the small glowing dots in the night sky, the stars, are actually suns like ours. They only appear dot-like to us because they are so far away. The distance of the sun is still given in kilometers, which is about 150 million. With stars, however, kilometers become unwieldy because of the size of the numbers. In popular science literature, the unit **light year** has become established for large distances, i.e. the distance that light travels in 1 year. This unit is convenient because it also tells us how far back in time we are looking when we look at an object at a given distance. In the technical literature, the unit **parsec** is used instead, which goes back to how the distance of closer stars is measured: Earth's position in winter is shifted 300 million kilometers compared to its position in summer, because it is on the other side of the Sun. Because of this change in perspective, the closer stars appear from Earth in a slightly offset position called a **parallax.** A parsec (from *parallax second*) is the distance at which a star appears offset by one arcsecond, an angle of 1/3600 degrees. This distance corresponds to about 3.26 light-years. A light year, in turn, is equivalent to 9.5 trillion kilometers, as mentioned earlier, which is about 63,000 times the Sun-Earth distance. Our nearest star, *Alpha Centauri,* is about 4 light years away.

These are already distances that seem huge to us. But it gets much better. In the mid-eighteenth century, it was discovered that the Milky Way, the whitish band that stretches across the night sky, is made up of billions of stars that form a massive structure, a **galaxy** of which we are a part and which we now know has a diameter of over 150,000 light years. Just a few years later, in 1755, Immanuel Kant expressed the hypothesis that some of the small nebulous patches that can be observed with a telescope are themselves galaxies like ours. This hypothesis was not finally confirmed until 1923, when Edwin Hubble (the same who later discovered the expansion of the universe) used a new super telescope near Los Angeles to resolve for the first time individual stars in our nearest neighbor, the **Andromeda Galaxy.** From the distance estimate, it appeared that these stars could not possibly belong to our Milky Way, so they had to form an independent system. The Andromeda Galaxy is about 2.5 million light years away from the Milky Way. Incidentally, it is currently moving towards us at 120 kilometres per second. In a few billion years, it will collide with the Milky Way – a cosmic super-event in which the two galaxies are expected to eventually merge into one. By then, Earth will probably no longer be habitable; so we'll have to find a better place to watch the spectacle.

Telescopes became better and better, and more and more galaxies could be observed and photographed. It is now estimated that there are several hundred billion galaxies in the **observable universe** (for the term, see below), which has a radius of about 45 billion light years, and each of these contains on average about 100 billion stars. A high proportion of these many stars will have planets, and trillions or quadrillions of these planets will have conditions similar to those on Earth. There is no doubt that life has evolved elsewhere.

How are distances in space actually measured? For objects that are only a few light years away, this is done with the help of parallax, as described above. For further distances, the measurement is indeed very difficult and only possible indirectly. Mostly it is a matter of **absolute** and **apparent brightness.** The absolute brightness says how much light the object emits in total. The apparent brightness tells us how much light from the object reaches us, i.e. how bright it appears to us. The distance can be determined from the ratio of the two quantities (the further away, the lower the apparent brightness compared to the absolute brightness). The apparent brightness is easy to measure. But how do you get the absolute one? The absolute brightness of stars or even galaxies can vary greatly.

In fact, there are only a few types of objects that are known from experience and theory to have uniform brightness: very specific classes of pulsating stars; very specific classes of stellar explosions, so-called type Ia supernovae. These objects are called **standard candles.** Finding such standard candles in other galaxies is therefore highly sought after. Otherwise, one often has to make do with somewhat cruder estimates by rule of thumb. In the past, distance estimates have sometimes been off by a factor of 10 or more. With all these difficulties, it is remarkable how accurately the universe has been measured in the meantime. Another method of determining the distance of very distant objects is redshift. However, because this has to do with the expansion of the universe, we will return to it only in the second point.

In general, most cosmological measurements are very difficult. After all, we are dealing with correlations that span distances of billions of light years and times of billions of years. Yet the theoretical part of cosmology is quite simple due to the high symmetry (uniformity) of the universe. The cosmological solutions are particularly simple solutions of GR. Remember the analogy with the Earth: geometry on a large scale is much simpler than geometry on a small scale. The Earth as a whole is, to a good approximation, a sphere, very symmetrical. If you know the circumference of the sphere, that is, a single number, then the sphere is already completely determined. Only if you look more closely at the small irregularities does it get complicated. To measure a mountain range in detail, to create an elevation profile, requires thousands of numbers, not just one. It's much the same with the universe. In the big picture, where we overlook the "little bumps" (galaxies and such), a handful of parameters are enough to characterize it. Only in the "small", when we describe the galaxies and stars, it gets complicated.

As simple as this may be in theory, however, it is difficult to measure these one handful of parameters that exist in cosmology. Usually one can only determine certain combinations of the parameters and must compare a number of different observations to extract the individual parameter values from the various combinations. In addition, further assumptions about standard candles and other astrophysical facts that are only roughly understood usually enter into the interpretation of the measured data. It was not until the 1990s that the era of precision cosmology began due to further improvements in observational technology, especially satellites with ever better telescopes and cameras. Before that, quantitative statements were usually associated with large uncertainties. Thus, cosmology is the subfield of fundamental physics that was the last to be able to deliver precise quantitative results.

Yet one crucial question remains completely unanswered: How big is the universe? In particular, is it infinite? The main problem here is the **cosmological horizon** that limits the **observable universe.** Since the universe has a finite age (more on that in a moment), any light since its creation could only travel a certain distance. We simply cannot look any further. The boundary thus set defines the cosmological horizon. Everything that lies beyond it is hidden from us, and thus also the answer to the question of whether the universe (1) ends somewhere, i.e. has a real boundary, or (2) is infinite or (3) is "closed", i.e. finite but boundless, like a spherical surface: anyone walking around the earth never comes to an end of the world, yet it is not infinite, but anyone who always walks straight ahead comes back to the starting point.

An answer could have been given by the question of spatial curvature. We remember from Sect. 7.4 that the cosmological solutions of GR allow us to separate space and time much more neatly than otherwise usual in GR or SR. Therefore, we can also quite clearly distinguish the curvature of space from the curvature of spacetime here. In cosmology, space is assumed to be uniform, more precisely: **homogeneous** (it looks the same in all places) and **isotropic** (it looks the same in all directions). Thus the geometry of space can be described by a single parameter, the curvature parameter. If this is positive, then the space is curved similarly to a spherical surface, but with one more dimension (a "hypersphere"). In particular, the space is finite. Geodesics in this space are circles, like the great circles on the spherical surface. "If you always walk straight ahead, you will come back to the starting point." If a positive spatial curvature were observed, we would know that answer 3 is the correct one. At the moment, however, there is nothing to suggest such a positive curvature.

If the spatial curvature parameter is zero, then the space is flat, that is, Euclidean. If the parameter is negative, the space is called hyperbolic. This means, as described in Sect. 7.4, that parallels diverge or that the sum of angles in a triangle is smaller than180°, both however only with measurable effects on the scale of billions of light years. The measurements from cosmology strongly suggest that space is Euclidean within the limits of our measurement precision, i.e. that the curvature parameter is zero.

In this case, the question of the size of the universe remains open. It could be that space is infinite (so answer 2). But it could also be that there is a very small positive curvature, below the limits of measurability. Then the universe would be very large, much larger than the observable universe, but not infinite (answer 3). Or there is a real boundary to space somewhere (answer 1). Our theories give no real reason why there should be such a boundary, but it cannot be ruled out. Certainly, however, there are more supporters of hypotheses that postulate a boundary between two different parts of space, parts in which, for example, different physical laws or different matter distributions prevail.

Or perhaps the space is indeed flat and unbounded, but still finite (answer 3), because the space is "topologically nontrivial". We recall the discussion of the torus in Sect. 7.4, where the torus was defined as a *flat* rectangle with opposite sides identified, meaning: Whoever runs out of the top of the rectangle comes back in from the bottom, and vice versa, and the same with right and left. Something similar could happen in the universe, with a cube instead of a rectangle. This possibility was first discussed by Karl Schwarzschild in 1900, even before GR existed.

Probably we will never know which of the possibilities is realized in our universe, because the answer is hidden far behind the horizon of observation. But perhaps we will find a fundamental theory that can be tested and which, in addition to other observable effects, also determines the size of the universe. One should never give up hope.

If the universe is infinite, that would be very interesting from a philosophical point of view. In an infinite universe there would be an infinite number of planets. On a small fraction of them, which is still infinite, life arises. On a tiny fraction of them, but still infinite, even intelligent life would arise. On a tiny fraction of it, but still infinite, people walk around looking just like us. In an infinite universe, all physically possible scenarios are realized, however improbable they may be. In an infinite universe, every possible story is a true story. All the stories you can make up, unless they contain physically impossible details, are actually happening somewhere right now. If you wish you had decided differently at some point in your life, then somewhere on some planet there is a copy of you running around whose story starts out just like yours, but who then, by some tiny deviation, decided the way you now wish you had. Better yet, there are an infinite number of such near-copies. For some, things go well after the decision; for others, everything goes wrong. It's best to think of the ones where everything goes wrong, then you'll know that your decision was the right one after all.

The Universe Is Even Getting Bigger

From QM we know that atoms have only very specific energy levels. We also know that the wavelength of light is inversely proportional to the energy of the photons of which it is composed. It follows that atoms can only absorb light at very specific wavelengths, namely those at which the photons have just enough energy to lift the atom from one level to another. Stars emit a continuous spectrum of wavelengths. On the way out, however, the light must pass through the star's atmosphere, where the wavelengths that match the atoms are absorbed. Therefore, if you split the light that arrives on Earth by wavelength (for example, through a glass prism), you will find numerous spots that remain dark, precisely at the wavelengths that were absorbed. Since all stellar atmospheres contain more or less the same elements, the dark spots are also the same in all stars; they form certain characteristic patterns.

For some stars, these patterns are slightly shifted, towards smaller ("blue shift") or larger ("red shift") wavelengths. This is explained by the famous Doppler effect. Stars move at individually different speeds, some moving crossways, some coming towards us, some moving away from us. When a star is coming towards us, the light wave appears somewhat "squeezed" to us, the light is blue-shifted. When it moves away from us, the light wave appears to us somewhat "pulled apart", the light is red-shifted. Since the directions are more or less random, the number of red-shifted stars is about the same as the number of blue-shifted stars.

With galaxies it is similar at first. After all, they also consist of stars, and their light therefore has the same pattern of dark spots when split up according to wavelengths. For the closer galaxies, the distribution is again random, some moving towards us, some moving away from us. The Andromeda galaxy, for example, is blue-shifted because it is moving toward us. But at greater distances, Hubble found a systematic relationship in 1929: all galaxies farther away were redshifted, and the redshift seemed to increase roughly proportionally with distance. This proportionality is known as **Hubble's law.** Hubble had discovered the expansion of the universe!

We remember: Expansion means that distances between fixed points in space (defined only because of the peculiarity of cosmological solutions) become larger, by a uniform factor. An expansion of 10% means that distances of 100 million light years grow to 110 million light years, and distances of 1 billion light years grow to 1.1 billion light years. It does *not* mean that matter expands out into empty space. Each galaxy more or less stays in place, except for small individual movements. We shouldn't imagine that distant galaxies are hurtling away from us due to expansion. Expansion has nothing to do with velocities in any real sense, so the speed of light does not form a limit. The distance between two galaxies can increase by 10 light years in a year, for example, without that being a contradiction. It is better to imagine that the increase in size is due to the creation of "new space" between the galaxies.

The redshift associated with expansion is therefore *not* due to the Doppler effect, as some books would have you believe, but to the fact that light waves are pulled apart by expansion on *their way* from the light source to us. The fact that galaxies twice as far away are redshifted about twice as much is not because they are moving away from us twice as fast, but because the universe had about twice as much time to expand while the light was on its way. Therefore, Hubble's law is only valid as long as the light travel times are small enough so that the expansion over this time can be considered uniform to a good approximation (over the entire history of the universe, it is not). More or less, we can say that Hubble's law is valid for distances that are (1) significantly larger than ten million light years, so that the redshift due to expansion outweighs the redshift due to the Doppler effect, and (2) significantly smaller than 10 billion light years, so that the expansion along the light path is reasonably uniform.

If you calculate the expansion back into the past, you find that at a certain point in time, about 14 billion years ago, all distances were zero. This hypothetical moment, when all space was reduced to a single point, is called the Big Bang. Underlying this expression is the idea of a great explosion. You are welcome to imagine a banging sound if that makes the Big Bang seem more real to you. What you are *not* supposed to imagine, please, is that at that time there was a tiny lump of matter in an otherwise empty space and that this lump then suddenly exploded into that space. Because space itself was only a single point at that time, and matter could not be anywhere else at all but there.

Remember the end of Sect. 7.4. There I tried to explain that it is a good idea to use the word "universe" for the whole of spacetime and not, as usual, only for space. Let us follow this suggestion for a moment. Then the universe has no age, because spacetime contains all of time from beginning to end. The big bang then marks something like the south pole of the universe. We can visualize this better if we ignore two space dimensions for the moment, as in Sect. 7.3. We imagine the remaining two-dimensional structure of time and one space dimension as having the time direction in the north-south direction and the space in the east-west direction. Space at a given time thus becomes a latitude, that is, a line at a constant distance from the South Pole. At the South Pole the latitude has zero length, because the South Pole is a single point. It is in this sense that it is to be understood that space has size zero in the Big Bang. As you go north from the South Pole, the latitudes become longer and longer, and the distances between the longitudes become greater and greater. That is, space is expanding.

By the way, from the South Pole you can only go north. The question of what lies south of the South Pole makes no sense. For the same reason, the question of what was before the Big Bang makes no sense. From the Big Bang, you can only go in the direction of the future.

Temporally, we are moving further and further away from the Big Bang as we ride the now toward the future. Scientifically, however, we are getting closer and closer to the Big Bang as our telescopes and theories get better and better. With telescopes, we see further and further into the past as we look farther and farther away, because the light then traveled a correspondingly longer distance to reach us. However, in this way we only get within about 400,000 years of the Big Bang, because before that the universe was opaque.

However, with the help of confirmed theories, we can get much closer, down to about a trillionth of a second. What happened before that, in the first trillionth of a second, there are some speculative (i.e. not confirmed) theories for which we have no evidence.

The closer we get to the Big Bang, the denser and hotter the matter swirling around in space becomes. In the big bang itself, it is infinitely dense and infinitely hot. In the first trillionth of a second, the temperature is higher than the highest energies we have ever produced with particle accelerators, and that is the reason why we have no confirmed theories for this time.

Did the Big Bang really exist as an event? The answer is: We do not know. The Big Bang is the point at which space is concentrated to a point within the framework of a certain confirmed theory, namely GR. Like almost all theories, GR has a certain range of validity. The closer we get to the big bang, the more extreme the conditions become, and at some point t_{min} we leave the validity range of GR. For the time before that we need another theory, presumably a QFT that includes gravity, and all theories of this kind are highly speculative so far. We therefore don't know how space and time behaved before t_{min}, or if space and time are even meaningful concepts there. The required theory possibly demands completely different concepts, from which the "conventional" space-time then "emerges" in the sense of a *reduction by replacement*. Therefore, the expression "one trillionth of a second after the Big Bang"actually means: one trillionth of a second after the time when the Big Bang would have taken place, if GR were still valid in this area. The Big Bang is a singular point in spacetime, if we extrapolate GR beyond the limits of its validity. We should not think of it as a real event. This extrapolation is extremely useful. The Big Bang is a point in an exact solution of GR that describes the real universe very well except for a tiny fraction of a second, and it would greatly complicate all notations if, for the sake of political correctness, we cut out that fraction of a second and stopped using the word "Big Bang".

When we talk about a trillionth of a second, it sounds very short, especially compared to cosmic time scales that span billions of years. We understand what is going on in all these billions of years, only we do not understand the first trillionth of a second; it sounds as if we understand almost everything. However, this is a misconception. Our units of time make sense to us because they are oriented to uniformly occurring processes. The earth always revolves at (almost) constant speed around itself and around the sun, so a day is always a day and a year is always a year. But near the Big Bang, matter becomes denser and hotter, particles collide much more frequently, there is no question of uniformity. If we identify the big bang with the time $t = 0$, then we can say very roughly that between $t = 0{,}001$ s and $t = 0{,}01$ s about as much happens as between $t = 0{,}01$ s $t = 0{,}01$ s and $t = 0{,}1$ s or between $t = 0{,}1$ s and $t = 1$ s and so on. In this case, the usual units of time do not give a very good measure of the actual ratios. A logarithmic measure of time would be better here, that is, a measure in which the distance between "one tenth of a second after the Big Bang"and "one second after the Big Bang"is the same as between "one second after the Big Bang"and "ten seconds after the Big Bang". With such a measure, it is the same *ratios* (here the ratio 1:10) that make up the same distance, rather than the same *differences as* usual. Such a measure would do more justice to what happened in the early universe.

However, in this way the big bang is an infinite distance away: A step towards the big bang by a logarithmic unit of measurement always means division by 10 in units of seconds. However often we divide by 10, we only get closer to the big bang, but never reach it, as in the story of Achilles and the tortoise. While the paradox of Achilles and the tortoise can be easily resolved, because the logarithmic measure is *not* the right one there (in units of seconds, Achilles quickly overtook the tortoise), in the case of the Big Bang, one can argue that it really is "infinitely far" away: Since the events in its vicinity are getting closer and closer together, it may be that an infinite number of "new" things happen or states are passed through before even the first trillionth of a second has passed. In that case, the amount of things we don't know about that trillionth of a second is infinitely greater than the amount of things in the billions of years after that we do understand.

With the help of the confirmed theories we can go back to the time "one trillionth of a second after the big bang". The state we find there represents a kind of **primordial soup,** whose composition and properties can be calculated back from the present state of the universe, but which must result causally from the still earlier time, whose history is unknown to us. Three questions in particular are open in this respect:

1. The primordial soup was apparently extremely evenly distributed, across regions which at that time were not in any causal contact, i.e. which could not influence each other in any way. How can this be explained?
2. From the confirmed theories, one would expect that matter and antimatter initially existed in equal amounts and within a very short time annihilated each other through collisions. In fact, however, a very large amount of matter remained, enough to form hundreds of billions of galaxies, each with 100 billion stars. But there is virtually no antimatter left at all; we have to create it artificially in particle accelerators. So there must have been

a significant surplus of matter in the primordial soup. How did this surplus come about?

3. For all known elementary particles, we know approximately in what quantity they are to be found in the primordial soup (more precisely: with what density, i.e. how many particles per cubic meter). We also know how high the density of matter is overall, and we know this via gravity (or spacetime curvature, in GR-speak), which is generated by matter and whose strength we can now measure from the speed of the expansion and the behaviour of galaxies. Here we find a discrepancy: the sum of all known elementary particles adds up to only about one-sixth of the total amount of matter. The remaining five-sixths must therefore exist in a still unknown form of matter. Because we cannot see this matter, but only know about it indirectly through its effect, gravity, it is called **dark matter.** What does this dark matter consist of?

For all three questions there are speculative theories that try to give an answer. But there are no firm findings. Especially the last question seems disturbing. For the first two questions relate to events at the beginning of the universe: How did the primordial soup become so uniform? Why didn't matter and antimatter annihilate completely? Everything that happened afterwards does not depend on what the answer to these questions is, what matters is only *that* the primordial soup was so uniform and *that there was* still so much matter left. The last question, however, relates to a form of matter that still exists today and plays a part in the evolution of the universe. Even more so, it is the *main component of* matter in terms of quantity.

If we don't understand dark matter, how can we claim to understand the history of the universe since a trillionth of a second after the Big Bang? Well, the point is that quantity is not necessarily a criterion for how much there is to say about a thing. The known elementary particles have a very rich structure, interacting with each other in various complicated ways to form interesting entities such as atomic nuclei, atoms and molecules, and on larger scales, stars and planets. They lead to the complex structures we observe in the universe today. Dark matter, however, seems to interact with us only through gravity. At least, all other interactions are so vanishingly small that, despite numerous attempts, it has not been possible to make it show any sign of detection in particle accelerators or other experiments. The effects via gravity, on the other hand, are quite simple and can be determined by only two pieces of information: their quantity and the fact that they are *cold*. This is why it is referred to as **Cold Dark Matter.** "Cold" here means that it does not hurtle through space at speeds close to the speed of light, but rather makes itself comfortable in the gravitational field of galaxies. With these two pieces of information, its effect on the history of the cosmos is already fully established. Further information about its nature would be of great interest to us, but has no direct relevance to the events that will be recounted in the following.

Starting from the primordial soup one trillionth of a second after the Big Bang, we can let the evolution of the universe unfold from that time until today before our theorists' eyes. What we get to see is the version of Genesis that physics has to offer. Here we restrict ourselves to a few particularly important events.

The first 3 min were impressively described by Steven Weinberg in his classic book *The first three Minutes*. The book dates from 1977, when dark matter was not yet known and there were no precision measurements of cosmological parameters. Nevertheless, little has changed qualitatively since then in the story that is told there. One important event from that era is **nucleosynthesis,** after the temperature has dropped below a billion degrees: a certain fraction of protons and neutrons join together to form small atomic nuclei, such as deuterium (one proton and one neutron) or helium (two protons and two neutrons). The processes that do this are similar to chemical processes, except that in chemistry, whole atoms join together, whereas here it is the much smaller protons and neutrons. Therefore, as mentioned earlier, the amount of energy involved is more than 100,000 times greater than in chemical reactions, and that is why these processes take place at such high temperatures. Stable atoms cannot yet form at this temperature. Electrons and atomic nuclei are flung back and forth too much by constant collisions.

After about 400,000 years, the temperature has dropped to about 4000 K. Now atoms can form, the strength and frequency of the collisions are no longer sufficient to tear a large part of them apart again. This has an effect on photons in particular. Since they interact with all charged particles, having negatively charged electrons and positively charged atomic nuclei buzzing around separately was a problem for them. They never got very far, always being immediately absorbed or scattered by the next best electron or atomic nucleus. Atoms, on the other hand, are electrically neutral to the outside. Photons could now fly freely between them. At the same time, the density also decreased due to expansion to such an extent that a collision with an atom became less and less likely. The universe became transparent. Photons from that time are still flying. Expansion has since caused their wavelengths to increase by more than a thousand times, reducing their energy. It now corresponds to a temperature of a little less than 3 K. Electromagnetic radiation in this energy range is called **microwaves.**

These microwaves from the early days of the universe come uniformly from all directions, forming the **cosmic background radiation,** which was accidentally discovered in 1964 by the Americans Penzias and Wilson while making measurements for an entirely different purpose. (In *The First Three Minutes,* Weinberg wonders why it wasn't systematically looked for much earlier). The discovery of background radiation was considered definitive proof of the Big Bang hypothesis.

The background radiation is an invaluable relic from the early universe; it can be used for excellent cosmic archaeology. Of particular value is the distribution of its temperature (i.e., its mean wavelength) as a function of the direction from which it comes. For example, it is found that the radiation in a certain direction of the sky is about one permille "warmer" than that coming from the opposite direction. In other words, one side is slightly blue-shifted, the other slightly red-shifted.

In Sect. 7.4 I said that in the cosmological solutions of GR one can again define a reference system which is in a certain sense "absolutely at rest", in contrast to the statements of SR. This system of rest is precisely the system in which the background radiation appears uniform. The red or blue shift of the background radiation is due to the Doppler effect, relative to this rest system. This allows us to determine the *absolute velocity of* the Earth through the cosmos: It is about 370 kilometers per second. The background radiation thus provides us with the exact opposite of the Michelson-Morley experiment. This experiment sought to measure the absolute velocity of the Earth in space based on the deviation of the speed of light and concluded that there are no such deviations, leading Einstein to conclude that all uniformly moving systems are equivalent and that there is no such thing as absolute velocity. With the measurement of the background radiation, however, an absolute velocity can now be determined after all. The difference is that Michelson and Morley started from a hypothetical medium called the "ether", which should fill the entire universe and represent the absolute rest system of electromagnetism, especially of light propagation, and thus *any* light is apt to measure velocity relative to this rest system. The crux, however, is that only a very specific light, namely the background radiation, defines such a rest system, and therefore only measurements on the background radiation can determine our absolute velocity.

SR applies "locally", i.e. on length and time scales small enough that the curvature of spacetime need not be taken into account, just as we can safely neglect the curvature of the earth's surface in a city map. There is no absolute velocity there, all uniformly moving reference frames are equivalent. With respect to the cosmos as a whole, however, which has a certain geometric structure that deviates from that of SR, there is an absolute system of rest.

The background radiation originates from a certain time, namely 400,000 years after the Big Bang. It has therefore travelled a very specific distance on its way to us. This distance is the same in all directions and defines the radius of a spherical surface. For us, this surface is the boundary of the observable universe. Beyond it, the universe is opaque (or it *was* opaque at the time it should have emitted a beam of light that would have had enough time to reach us). The light from the background radiation has traveled about 14 billion light years on its way to us. However, each piece of that path has continued to expand after the light has traversed it. Therefore, the current distance of the area from which the radiation originated is about 45 billion light years. In particular, the surface has moved 45 billion light years away from us since the Big Bang, within 14 billion years – evidence that the speed of light is not a limit in cosmic expansion.

The directional temperature difference of 1 permille, which is caused by the motion of the Earth in space, has the form of a "dipole": it consists of a contrast between opposite directions, blue-shift in the direction of motion and red-shift in the opposite direction. However, there are also small variations in the temperature of the background radiation in other directions, i.e. minimally blue- or red-shifted spots. The structure and distribution of these spots give us valuable clues about the composition of matter in the early universe and are therefore ideally suited to test our theories and reduce the inaccuracies in our knowledge of cosmological parameters. Therefore, the measurement of these fluctuations has been one of the main concerns of precision cosmology in the last few decades.

The spots are largely due to small density fluctuations in the early universe. Basically, therefore, it is an **acoustic spectrum.** The sound of a musical instrument is produced by its vibrations, which cause pressure fluctuations in the air that propagate as *sound.* Each instrument has its own specific sound. This is because each instrument, when producing a certain tone, also produces some *harmonics,* that is, tones with a multiple of the frequency of the fundamental tone, in a certain ratio characteristic of the instrument. We can think of density fluctuations in the early universe as being similar to this. Depending on how they originated and in what composition of matter they propagate, a characteristic spectrum emerges. In measuring the tiny irregularities of the background radiation, one has indeed been able to identify specific wavelengths of "sound"in the early universe and their harmonics.

The small density fluctuations in the early universe are also the seeds for the formation of stars and galaxies. Gravity has the property of amplifying such fluctuations. Dense regions attract further matter by means of gravity and thus become even denser (just as unregulated capitalism is said to make the rich richer and the poor poorer). This process is called **structure formation** in cosmologists' jargon. The "agglutination" of the universe continues for about 400 million years without any special events. Then the first galaxies are formed, clusters of matter whose density is so much higher than that of their surroundings that they decouple from the expansion. Galaxies no longer expand. Within the galaxies, clouds of gas form and collapse under their own weight, heating up in the process. This is how the first stars are formed. They shine, replacing the background radiation, which has now cooled and thinned out considerably, as the dominant source of light in the universe.

At that time, matter was almost exclusively hydrogen and helium. Nuclear fusion takes place inside the stars, creating some higher elements. At some point, stars run out of fuel and enter a dramatic final phase of their lives, leading in many cases to a **supernova,** a gigantic explosion. In this final phase, all the higher elements are also created in cascades of nuclear fusion reactions and ejected into space with the supernova: the matter of which we are made. For a second generation of stars and planetary systems could now use these elements, planets with solid surfaces could be formed, water could be formed from hydrogen and oxygen, and together with carbon, complicated molecules could be assembled. So a whole generation of stars had to explode to make us possible. Let's face it, isn't that a very romantic thought?

The Universe Is Even Getting Bigger Faster and Faster

How will the history of the universe continue in the future? Until the 1990s, people were pretty sure that gravity would slow down the expansion in the long run. So there were two scenarios, much like firing a rocket: 1) If the rocket is sufficiently fast as it flies up, it will be slowed by the Earth's gravitational field, but not enough to keep it from leaving the Earth's gravitational field and flying out into space forever. 2) The rocket is too slow, the slowing by gravity brings it to a halt at a certain altitude, and from there it plummets back down to Earth. It's the same in cosmology: either the universe keeps expanding despite slowing down, or the expansion comes to a halt at some point and runs backwards from there until the universe finally ends as it began: in a single point, a kind of "north pole" as a counterpart to the "south pole" of the Big Bang. This idea appeals to many romantics, it is so beautifully symmetrical, and it also opens up the possibility of a cyclical universe, because who knows, maybe the end point (called the *Big Crunch,* as a counterpart to the *Big Bang*) is at the same time a new Big Bang, and the whole story starts all over again.

But it doesn't look like that at all. Measurements in the 1990s showed that the expansion only slowed down in the early days of the universe, but is now *accelerating*. This would not be possible with Newtonian gravity. But GR allows forms of energy that drive everything apart instead of pulling it together. The cause of this accelerated expansion is called **dark energy**, not to be confused with dark matter, which is attracted by gravity.

Dark energy has the remarkable property that its density does not decrease with the expansion of the universe, but always remains the same. If the expansion has increased all distances by a factor of 2, then the density of normal matter has decreased by a factor of 8: After all, the same amount of particles is now in a volume eight times larger. To put it more delicately, the density of matter scales with the reciprocal of the third power of the scale factor. Since the density of dark energy does not change, this means that its ratio to the density of normal matter increases sharply. In the early universe, the density of dark energy was so small in proportion that it did not matter at all. In the primordial soup, we need not consider it. Today, however, it represents about 70% of the total energy of the universe, and the ratio continues to increase. The fact that it has dominated the universe for only a few billion years also explains why the acceleration of the expansion has only recently begun.

Dark energy finds its way into GR in two ways: 1) as a term which can simply be added to Einstein's field equations without affecting their consistency. This term has the name **cosmological constant** and was already proposed by Einstein, but subsequently discarded. Now it seems to be needed again after all. We would write it on the *left-hand* side of the equations to denote that it does not arise from the properties of matter, which are known to be on the right-hand side of the equations by convention. 2) As ground state or **vacuum energy** from QFT, which has the same properties as a cosmological constant with respect to gravity. We recall that vacuum in QFT is the state of *minimal* energy, but this minimal energy can in principle be arbitrarily high. This vacuum energy provides a contribution to the *right* side of Einstein's field equations, because it is connected with the properties of matter.

If one combines the cosmological constant with the vacuum energy, the result is the total amount of dark energy. So far the theory. The problem is that this vacuum energy, as it comes out in the calculations of QFT, is in principle *infinite* and we do not know what to do with it. Therefore, we probably still have to better understand the interplay between GR and QFT, i.e. quantum gravity, in order to be able to classify how a dark energy of the magnitude we observe comes about.

With dark energy, it looks like the universe will expand for all time, and faster and faster. The galaxies, as I said, are decoupled from the expansion, meaning that they are moving away from each other (or groups or clusters of them are moving away from each other), but within the galaxies the distances remain as they are. At some point, a few trillion years from now, no new stars can form because the fuel, hydrogen and helium, is used up. The last stars burn up. And from then on, the universe is completely dark, cold, and most likely devoid of life. (But who knows, we know the slogan "Life finds a way").

The accelerated expansion has another effect: it aggravates the problem of horizons. We have already talked about the **cosmological horizon**: since the Big Bang, light has only been able to travel a certain distance, so we cannot look as far as we like. However, this is initially just a matter of waiting. If the universe does *not* expand at an accelerated rate, then every event, i.e. every point in space-time, will at some point lie within our observation horizon. This is because the horizon is moving away from us as more and more time passes that the light has available for its journey. In this calculation, however, we must take into account that the expansion also causes the distance that the light has to travel to increase during its journey. For uniform or slowed expansion, however, light always manages in the end, no matter when or where it was emitted (black holes aside), to reach every point in space at some point, including us. So all we have to do is wait. We may have to wait a very long time – billions, trillions, quadrillions of years – but eventually any distance will be bridged, assuming our telescopes are good enough. If the universe should be closed (in the sense of "If you always go straight ahead, you will come back to the starting point"), then at some point on the horizon we will catch sight of ourselves, more precisely our past: The light has traveled once around the entire universe.

With accelerated expansion, this is no longer true. Distances increase too much during the journey of light. A beam of light emitted today in our direction at a distance of a little more than 15 billion light years will *never* reach us. An **event horizon** has been added to the cosmological horizon: Events whose distance exceeds a certain value can never make contact with each other. As a result, distant regions of the universe are forever isolated from each other. Even eternal waiting and the best technology can no longer help us here.

8

The Unknown

8.1 The Hunt for the Theory of Everything

The work on the Standard Model of particle physics had reached a stage in the mid-1970s where the fundamental questions seemed to be essentially solved. Two theories covered the entire field of particle physics with great accuracy, at least on the energy scales accessible to experiments: first, **QCD,** the QFT of the strong nuclear force; second, the **electroweak theory,** i.e., the QFT of the electromagnetic force and the weak nuclear force. A third theory very accurately described gravity, a force that plays no role in particle physics but dominates in the interaction of large amounts of matter: **GR**. Together, these three theories seemed to describe everything we find in the universe, from the smallest particles to the largest clusters of galaxies and the expansion of the universe, from the first trillionth of a second after the Big Bang to a trillion years in the future.

Physics could thus look back on a gigantic success story. Since the "big bang" of modern physics, which was triggered by Newton's *Principia* in 1687, things have progressed steadily. Since the middle of the nineteenth century, development has been accelerated even further. Maxwell's theory of the 1860s became the prototype of a unified field theory that reduced very many phenomena to very few equations. Thermodynamics and Statistical Mechanics described everything that had to do with heat, and brought the bustle of the smallest particles into play. SR and GR of 1905 and 1916, respectively, explored the structure of space and time. During the first half of the twentieth century, the structure of atoms and their building blocks was revealed, and with QM a suitable language for describing them was found. The reactions of

© Springer-Verlag GmbH Germany, part of Springer Nature 2022
J.-M. Schwindt, *Universe Without Things*,
https://doi.org/10.1007/978-3-662-65426-2_8

the atomic shells defined chemistry, the reactions of the atomic nuclei explained radioactivity and brought to light the tremendous energies of nuclear fission and nuclear fusion. This period also saw the discovery of the expansion of the universe, thus establishing the Big Bang model as the standard model of cosmology. In the 30 years following the Second World War, the world of elementary particles was further explored, the particle zoo was catalogued and their interactions were measured with precision to many decimal places. And so now one had arrived at three theories, at a few equations that encompassed all of this. Two of the theories, QCD and the electroweak theory, are quantum field theories and therefore very similar to each other, so that they can also be regarded as two parts of a single theory, the Standard Model of particle physics.

Spurred on by the successes, many physicists at that time, i.e. around 1975, were convinced that physics could be brought to a conclusion within a few years or perhaps decades at most. Many younger particle physicists already feared that they would soon be out of a job. There was great optimism, and the search for the all-encompassing theory, the **Theory of Everything,** began. Einstein had already spent the last decades of his life unsuccessfully working on such a project, but now the mood was such that *everyone was* rushing into it; it was *the* one great project of fundamental physics, everything else was secondary.

The first attempt was to further unify QCD and electroweak theory, to represent them as two variants of a single force, in much the same way as electromagnetic and weak nuclear forces had previously been unified into electroweak theory. Here, for the first time, physics was carried out according to the **modular principle.** In the description of the forces, certain mathematical structures, the so-called *Lie groups,* play a major role. These are fully classified (as discussed in Chap. 3: Mathematics is the zoology of mathematical structures). Each force is represented by a Lie group, which there takes on the role of a "gauge group", i.e. a charge scheme (Sect. 7.8). To unify two forces, one must represent their Lie groups as subgroups of a larger Lie group. There are many ways to do this, and, according to the classification, one tries the simplest ones first and, if unsuccessful, works one's way to the more complicated ones. In addition, the particle zoo must also fit these Lie groups in a certain way, i.e., the physical zoo of particles must be mapped into the mathematical zoo of representations of Lie groups ("representation" here is a mathematical term that would be too complicated to define here). Tinkering with these theories has a certain resemblance to handling toy building blocks that are to be put together to form a certain overall picture.

These *Grand Unified Theories* (GUTs), as they are called, are not to be confused with the theory of everything, which is supposed to include gravity as well. The GUTs, in whatever guise, do not predict anything different from QCD and electroweak theory for the energy scales at which we experiment with particles. Only at a much higher energy scale does the unification become apparent.

Energies in particle physics are usually expressed in **electron volts** (eV), which is the kinetic energy that an electron absorbs when accelerated by a voltage of 1 volt. One billion electron volts is one gigaelectron volt (GeV). The most powerful particle accelerator of our time, the LHC, accelerates particles to energies of up to about 14,000 GeV. The scale at which GUTs become relevant is about 10^{16} GeV. Particles at such energies would have to be shot at each other to experimentally test the GUTs. Unfortunately, this energy is a trillion times higher than even the LHC can produce.

Fortunately, however, there is also a prediction that is realistically observable to us: proton decay. The GUTs predict that protons have a certain half-life in which they decay into other particles, but a very large half-life of over 10^{30} years (1000 billion billion billion). But since we are surrounded by so many protons, one of them would have to decay from time to time. Registering such individual decays is not easy, but experiments to that effect have been done. Unfortunately, no decays have been found, from which it follows that the half-life is at least greater than the simplest GUTs predict. Thus it remains unclear whether the chosen approach to unification was correct.

The other big construction site is the inclusion of gravity. One can understand the *metric*, i.e. the quantity which determines the geometry of spacetime and which is responsible for gravity according to GR, as a field. Thus, GR is a classical field theory, which one would like to extend to a QFT. But it turns out that the methods which have served for this extension in the case of the other forces, in particular the method of Feynman diagrams, do not work in the case of gravitation. So to make a QFT out of GR, you need other methods and a deeper understanding of QFT. But maybe it doesn't work at all. Then one would need another, new theory, which is still below GR and QFT in the hierarchy of theories and from which the two can be derived.

One of the two possibilities, i.e. either to define a QFT of GR or to reduce the two theories together to a new theory, is necessary for reasons of consistency. For at present we have no idea how the behaviour of quantum objects feeds back into gravity. Any form of energy generates a gravitational field. Now, if a quantum object is "a little bit here and a little bit there" in the sense of a quantum mechanical superposition and then shrinks down to a location in the course of a measurement, then the same would have to happen to the gravitational field (or spacetime curvature) generated by the quantum object. At the same time, however, gravity (or spacetime curvature) has a feedback effect on the motion of the quantum object (and on other quantum objects in the immediate vicinity); this is an interaction, and it is precisely for this that we do not yet have a consistent description. At the energy scales we usually work with or can produce in particle accelerators, the expected effects are extremely small, well below the threshold of measurability. But for large energies on the **Planck scale,** roughly 10^{19} GeV, which is still about 1000 times more than the GUT scale, the effects become relevant. So we already *know* that we have no valid theory that can successfully describe phenomena at these energies. We know this even though we have never done experiments in this range and probably *never* will be able to do such experiments. But we can think up such experiments and have no idea what would happen *if* we did them.

How is such a problem to be evaluated? Two theories that are inconsistent with each other in their present state, but the inconsistency only affects phenomena that we may never be able to observe? I think almost all physicists are naturalists at least to such a degree that they assume that "nature" must have somehow solved the problem, that a solution *exists* that describes the "world" (whatever the world may be in this new theory then) at this scale. Even if we cannot observe processes of this kind for practical reasons, they are possible in "nature" and most likely did occur in the early universe. Inside black holes (where we unfortunately can't look either!) they probably take place as well.

On the ambitious way to the theory of everything, the problem had to be solved in any case, and so a whole generation of theoretical physicists (experimental physicists could not do anything here, after all) threw themselves into it. The hope was to arrive at a theory that was so clear, unambiguous and "elegant" that it simply *had to be* right. This pipe dream did not come true. A number of highly speculative theories were put forward and developed over a number of decades, but without any clear or even proven consistent result. Gradually, over the course of the last 40 years, one could observe how the initial optimism of the theory of everything hunters declined and slowly, step by step, gave way to frustration as well as a certain disorientation. For about 20 years now, there has been talk of a "crisis".

The problems had been underestimated that arise from the fact that the answers to crucial questions are to be expected on an energy scale to which we have no experimental access, so that the theoreticians had to tackle the work single-handedly. Only once in the history of physics has a successful theory been established without any experimental input, and that was with Einstein's GR – unquestionably a brilliant intellectual achievement. In all other cases, groundbreaking observations either preceded the theories, or the development of the theory went hand in hand with corresponding experiments. A simple physical "principle" leading us to the theory of everything, as the equivalence principle led Einstein to GR, is not in sight. Instead, the majority of attempts have followed a building-block methodology, in which building blocks and mathematical "recipes" of known theories have been generalized and puzzled together anew, but without success. A theory of everything seems a long way off.

While theoretical particle physics got bogged down in its all-encompassing speculations, experimental particle physics made quite a lot of progress. Ever larger, more powerful particle accelerators were built, in which particles were shot at each other on precisely controlled trajectories at ever higher energies. The latest and most advanced of these is, as mentioned, the LHC (Large Hadron Collider) at the CERN research centre near Geneva. Located in a 26.7 km underground ring-shaped tunnel, it has been in operation since 2008 and shoots protons at energies of up to 14,000 GeV at each other. It is an engineering feat, one of the greatest of our time. Planning and construction took about 14 years, involving more than 200 research institutes from around the world, and a total of more than 10,000 scientists, engineers and technicians. The engineering advances that have been made are immense and, in the opinion of many, far outweigh the scientific value to particle physics. Of course, there has been a search for effects beyond the Standard Model that might provide any clues as to how particle physics might proceed. However, nothing of the sort has been found, only the Standard Model confirmed over and over again. After all, the Higgs particle was discovered in 2012, the last building block of the Standard Model that still had to be proven.

The good news is that the Standard Model is extremely good, by far the best and most precise theory we have ever had. Before, it had always been the case that with each new order of magnitude on the energy scale, new, unexpected particles and new correlations had appeared. However, the Standard Model has survived several generations of particle accelerators over the last 40 years without needing to be extended (apart from a few very minor adjustments).

The bad news is: if this continues, particle physics as a research discipline is finished. At the moment, the successor to the LHC is being discussed; it is not yet clear whether, where and when it will be built. It will certainly cost many billions again, and in return it will produce an energy per particle that is perhaps ten times higher than that of the LHC. But what do we hope to learn from it? In the worst case, only that the Standard Model is still valid at this energy, as expected. At least, the interactions of the Higgs field could be measured a little more precisely – a detail in the parameters of the Standard Model.

Of course, it may also be that we are surprised with new physics that comes marching around the corner just at this energy, something that finally points us beyond the Standard Model and thus gives particle physics a new perspective. A tough bet to make here.

From a theoretical point of view, the evidence is that new physics can be expected in the range of the GUT scale and the Planck scale, between 10^{16} and 10^{19} GeV, far beyond what is experimentally feasible. Whether anything is already happening below that is uncertain.

While particle physics has become somewhat disoriented, we are still in a golden age in cosmology. Only in the last few decades have precision measurements of the parameters that describe the universe at large become possible. To everyone's astonishment, it turned out in the 1990s that only slightly less than 5% of the total energy in the observable universe is covered by the particles of the Standard Model. The large remainder is divided into two components with different properties, known by the nicknames **dark matter** and **dark energy**. Dark matter participates in galaxy formation and accounts for about 80% of their mass. Dark energy is completely uniformly distributed and drives the accelerated expansion of the universe. This gives physics two further tasks, namely to better understand these two components, which only play a role on galactic length scales and therefore do not appear in experiments on Earth.

Doing cosmology is a very different kind of science than doing particle physics. We can neither influence the events that extend over distances of billions of light years, nor reproduce them on Earth. Therefore, there are only observations in this field, no experiments in the true sense. On the theory side, things also look quite different. In cosmology, all physical theories play together. The ratios of matter constituents in the universe today come from particle reactions in the early universe, so particle physics is in play. Since there are a lot of particles involved, statistical mechanics is needed. Stars and also galaxies shine because of nuclear fusion – a jurisdiction of nuclear physics. Gravity, which makes galaxies form and holds them together, is described by GR, as is the expansion of the universe. The strong magnetic fields that are found in the vicinity of some cosmic objects follow Maxwell's theory.

This is a problem when it comes to unknown phenomena. Because which theory is now responsible? Something corresponding to dark energy occurs in both GR ("cosmological constant") and QFT ("vacuum energy"). But how do the two interact, and might something else need to be added? Is dark matter unknown elementary particles? Or does it perhaps consist of black holes described by GR? Or does even the theory of gravity have to be modified on large length scales? The many possibilities do not make it easy to find the right solution.

Overall, cosmology is somewhat later than particle physics in terms of its findings and precision measurements. However, here, too, there are indications of developments that correspond to those in particle physics and are beginning to put the expression "golden age" into perspective. There is now also a standard model of cosmology that describes the development and composition of the universe. Essential parts of this model already existed in the 1970s, but at that time all parameters were still subject to very large uncertainties, while particle physics was already fixed to many decimal places, and nothing was yet known about dark matter and dark energy. In the meantime, most parameters have been measured with satisfactory accuracy. The measurements are becoming more and more precise, the universe is being illuminated more and more completely, and the significance of every tiny spot of temperature fluctuation in the background radiation is being recorded, but little has changed in the overall picture. The abundance of data is yielding fewer and fewer significant insights. The Standard Model continues to be confirmed here as well. And the big open questions remain unsolved here too.

This essentially summarizes the situation in which fundamental physics finds itself today. We now want to assess the questions which have remained open in this process in more detail.

8.2   Open Questions

The views on what the great open questions of physics are differ to a certain extent, especially in the formulations. This shows that the tasks of physics and the meanings of the existing theories are understood differently.

Smolin's List

I would like to start with the list of the "five big problems of theoretical physics" from the book *The Trouble with Physics* by Lee Smolin (2006), in order to then assess this selection and the formulation of the problems from my own point of view:

1. *Combine general relativity and quantum theory into a single theory that can claim to be the complete theory of nature.*
2. *Resolve the problems in the foundations of quantum mechanics, either by making sense of the theory as it stands, or by inventing a new theory that does make sense.*

3. *Determine whether or not the various elementary particles and forces can be unified into a theory that explains them all as manifestations of a single fundamental entity.*

4. *Explain how the values of the free constants in the Standard Model of particle physics were chosen in nature.*

5. *Explain dark matter and dark energy. Or, if they do not exist, determine how and why gravity is modified on large scales. More generally, explain why the constants of the standard model of cosmology, including the dark energy, have the values they do.*

(Smolin 2006, p. 5 ff.)

Let's go through the problems in order:

Problem 1 The search for a quantum gravity is indeed a problem recognized by almost all physicists. First of all, it is a consistency problem: It is already very difficult to "write down" a self-contained theory at all, which is consistent with gravity and QFT. If this finally succeeds, it still remains to be seen whether it also describes the "real world", i.e. whether it can be confirmed by experiments.

For quantum gravity there are essentially two possibilities: (1) It turns out that a QFT of gravity is definable. It is known that this must be done in a different way than for the other forces; the method of Feynman graphs cannot be transferred. But there is some evidence, based on calculations with renormalization group methods, that a meaningful QFT of gravity is nevertheless possible (the related keyword is Asymptotic Safety; if you're interested, do some research on what that means). I think we still need to develop a much better understanding of QFT to finally decide the question. QFT as it stands today is optimized for scattering experiments, where particles are shot at each other in a vacuum, and for decays of unstable particles. A deeper mathematical understanding of QFT in more general cases is yet to come. Already in the case of bound states like the hydrogen atom (which after all was one of the simplest problems in the framework of QM!) we have difficulties to describe them in the framework of QFT. (2) But maybe in the end it turns out that a QFT of gravity is not possible. In that case, a new theory has to be found that reconciles both.

I find the second part of the sentence very daring, where it says that quantum gravity is supposed to be the "complete theory of nature". First of all, it is conceivable that the unification of gravity and QFT can be accomplished without involving the entire Standard Model of particle physics. It doesn't have to be a theory of everything that comes out of it. Besides, there may be other phenomena in the universe that are not accessible to our current experiments and that go beyond the scope of both GR and QFT. We simply don't know. So I would leave the "complete theory of nature" out of it here for now.

What are the chances that we will solve the problem of quantum gravity? The energy scale at which the effects of such a theory are expected is, as I said, the *Planck scale*, 10^{19} GeV, many orders of magnitude beyond what is experimentally accessible to us. With luck, however, it is conceivable that a theory will also leave some trace in the range we can work with. Whether such traces are then unambiguous enough to be considered clear evidence for the correctness of a theory is yet another question.

If a QFT of gravity turns out to be feasible from the theory side, then the theory that emerges is definitely a plausible working hypothesis. The consistency problem would then already be solved. The theoretical framework, QFT, would be the same as for the other forces, nothing new would be added, and no new, unusual hypotheses would be proposed. According to the principle of Occam's razor, one would probably interpret the circumstantial evidence for such a theory relatively favorably.

If, on the other hand, a QFT of gravity is *not* possible, then a completely new theoretical framework must be built, which is pure speculation to begin with. Certainly, it would already be a great success if such a framework were built in a consistent way. But in such a case, a much larger body of experimental evidence would be needed before the theory could be accepted with a good scientist's conscience. Whether such evidence can be mustered remains questionable, given the unattainable Planck scale. So it may very well be that quantum gravity fails because of the *practical limits of* physics, or more precisely, the practical impossibility of testing the theory at the scale where it becomes relevant. However, it is still too early to give up. Physicists are resourceful people.

Problem 2 Smolin, like any reasonable person, is disturbed by the mockery of all intuition in quantum theory. (By the way, here QM and QFT are meant together, the strange quantum behaviour prevails in both). One may ask, however, whether this is a scientific problem. The answer to this question depends on what claim one has to the natural sciences. That claim is a *philosophical* stance. Smolin emphasizes that he is a *naturalist*. He believes, first, in a world whose properties exist independently of the ways in which we humans interrogate them. Second, he demands that physics relate to this world and its properties, describe them, even explain them, and not just make statistical predictions with opaque but working tools, as QM in its present form unfortunately does. In doing so, he contradicts Kant, who repeatedly emphasized that the laws of natural science refer only to phenomena as they present themselves to us, and in no way to reality as it exists independently and outside of us. He also contradicts Duhem's view that a physical theory *explains* nothing (Chap. 6). If one agrees with Smolin in this philosophical demand on natural science, then QM is a problem that needs to be solved, and a thoroughly difficult one at that, on which the community of physicists has been racking its brains for 100 years. There are indeed naturalistic interpretations of QM, such as the many-worlds interpretation described above (and people who hold them, i.e. who think that a solution to the problem has already been found), but Smolin does not think that these are sufficient. As you know from Sect. 7.6, this is also my opinion.

As far as naturalism is concerned, I'm with Kant and Duhem rather than Smolin, and so in my view QM is not a problem that *needs to be* solved scientifically. Personally, I think the whole thing is more of a philosophical problem, so would place it beyond the *fundamental limits of* physics. But it *may* of course happen that QM and QFT will eventually be relegated to a more fundamental theory that makes more sense. I can certainly see value in QM in its present form. It almost seems as if the good Lord (please understand only metaphorically) allowed himself a profound joke with it, a kind of *memento mori* to natural science. He forces us to think about it, shows a variant of what science can be and what it can still leave open, even if everything numerical has been said.

Problem 3 The unification of elementary particles and forces into a single fundamental entity is a great pipe dream. A single field, a single force, a single theory, a single principle, a single equation, and everything else emerges from it, that would be so beautiful. The unified theory, if it is found, is so clear, so elegant, so unambiguous, that it simply *must be* true. This pipe dream has been haunting theoretical physics for several decades, afflicting the greatest of physicists, starting with Einstein.

Sarcasm aside, of course attempting further unification is an important task in physics. The existing experimental findings of particle physics should (again according to Occam's razor) be combined into one theory as efficiently as possible, and for this purpose it would certainly be better to work with only one field than with the extensive particle zoo of the Standard Model. Inspired by the successes achieved so far, it is also no wonder that this task was approached with great optimism (which, however, has gradually waned).

The question remains how clear and unambiguous such a theory would actually be, and whether we can test it experimentally. If the theory summarizes the existing experimental findings of particle physics more compactly than the Standard Model and also makes no other predictions that contradict the experimental findings, then one would already accept the theory without further testing. This is because it describes the findings more efficiently than the previous theory and is therefore preferable according to Occam's razor. However, this is not necessarily the most realistic scenario. In the experiments from the 1970s according to the modular principle, which I have described above, and the later experiments of the so-called *string theory,* it looks more like this: The unity of forces and particles only becomes apparent at a very high energy scale (the Planck scale or somewhat below), which is not experimentally accessible to us. Below that, various symmetry-breaking *mechanisms* operate to divide apart the various "manifestations" of the "unified field" (or whatever the fundamental object is in the unified theory), leading to the Standard Model of particle physics as an *effective theory* on the scales we can access. There are a variety of possibilities for these mechanisms. In most cases, whole cascades of additional particles and other phenomena arise (further manifestations of the fundamental object or its effects), which do not occur in the Standard Model and for which one must then partly think up new mechanisms again, which explain why they had to remain hidden from the eyes of the experimenters until now. The result is then in each case a theoretical construct, which may be incredibly beautiful and simple on the unification scale not accessible to us, but in the description of the scales accessible to us is much more ambiguous and complicated than the Standard Model. Such theories would need clear additional experimental evidence to raise them above the status of pure speculation. It may well be, therefore, that even if unification succeeds in theory, it will remain forever unproven because it fails due to the *practical limitations of* our experimentation – a situation similar to that in quantum gravity.

It remains to emphasize that the unification of forces and particles has no real physical problem to solve, unlike quantum gravity. There it is a matter of ironing out a real inconsistency between GR and QFT. But here we are only concerned with a *possibility to* achieve further unification. There is no inconsistency in the existing theories, no unexplained experimental data that makes this a *necessity.*

Problem 4 Why are the parameter values in the Standard Model the way they are? This question can be judged in two ways. First, one can say that theories contain certain constants: the gravitational constant, the elementary charge, the masses of the particles, and so on. The task of the natural sciences is to describe *how* things behave (at this point it doesn't matter whether we approach this task as instrumentalists or naturalists). *Why* a particular theory with particular parameter values describes our world is a question outside the *fundamental limits of* natural science. On the other hand, one *can of* course try to find a more fundamental theory to which the previous theory can be reduced, one that makes do with fewer parameters from which the more numerous parameters of the previous theory can be derived.

Thus the starting position is similar to problem 3: There is no *real* physical problem to be solved here, only a *possibility* that should be tested with theoretical and, as far as feasible, experimental methods, but which may also turn out to be impossible or at least not provable without losing anything crucial. A plausible conjecture is that problems 3 and 4 are closely related. Unification of fields and particles would certainly shed new light on the parameters of the Standard Model. Conversely, it seems realistic that an explanation of the parameters would also include a hint in the direction of unification.

If the explanation of the parameter values does not succeed within the framework of physical theories, two *metaphysical* explanation patterns are already available. It is the case that some of the parameter values seem to be virtually tuned to enable complex chemistry and thus life. This becomes apparent to a particularly astonishing degree when the parameters of particle physics and cosmology are combined. A great many things had to come together for the universe to build up this wealth of structure. This makes it easy to get the idea that the whole thing is personal; the question of parameters takes on an **anthropic** tinge: why are the parameter values such that human life is possible? The two metaphysical explanations then are:

1. The universe was created by a higher intelligence (a god or a civilization of aliens, depending on your taste) for the purpose of making intelligent life possible.
2. There is a multitude of universes – a so-called **multiverse** – described by different theories, or by the same theory but with different parameter val-

ues, and we logically find ourselves in one such specimen in which human life is possible.

The multiverse hypothesis is currently hotly debated, in particular whether it might not, at least under certain circumstances, be investigated within the framework of physical theories and thus acquire a scientific character. I will come back to this in Sect. 8.4.

Problem 5 Here Smolin lumps all cosmological problems together. I would rather separate them from each other.

Problem 5a What is dark matter? For a change, this is once again a real physics problem that *needs to be* solved in order to further complete physics. As already described, it is relatively difficult to say which branch of physics is responsible for this phenomenon, which can essentially only be seen in the rotation speeds of the galaxies, which suggest a higher mass than could be explained by the known particles. Most assume that these are still unknown elementary particles, but other suggestions have been made. With the progress of cosmology, quite a few possibilities could also already be ruled out, according to the scheme: If dark matter were ..., then this would have also the effect, that ..., since this however contradicts the cosmological observation ... dark matter must be something else. However, there are still plenty of possibilities left. The case is particularly difficult if dark matter consists of particles that interact with Standard Model particles only via gravity. This would mean that we would not be able to "grasp" it in any way. A meteorite made of such matter could hit the Earth without us noticing, it would just whiz through us and come out the other side of the Earth without leaving the slightest trace. Therefore, dark matter may very well be a problem whose solution fails because of the *practical limits of* physics. If we cannot subject it to tests for lack of interactions, all that remains is to estimate its total amount from the motion of galaxies, as we have done so far.

Problem 5b What is dark energy? One often reads that much less is known about dark energy than about dark matter. But this is not correct. In fact, we have *two* theories that involve a form of energy that behaves exactly as we observe dark energy to behave: the cosmological constant in GR and the vacuum energy of QFT. The question is how the two combine to explain exactly the amount of dark energy we observe. (Perhaps a third, as yet unknown, thing also plays into it, but there is no evidence for that at present). Sorting out this interplay is a task for quantum gravity. Thus, Problem 5b is, in my view, a part of Problem 1.

Problem 5c How are the parameter values in the standard model of cosmology to be explained? The same applies here as in problem 4.

In summary, it follows (from my point of view; views differ widely) that only problems 1 and 5a are "real" physics problems that *must be* solved in any case before one can hypothesize that the foundations of physics are complete. (I am only saying that this would be necessary, not sufficient.) The others can be dispensed with if need be. For both problem 1 and problem 5a, there is a very realistic possibility that we will never solve them because of the *practical limits of* physics, i.e., the limits of what humanly possible experiments can do. Of course, one should still try, even if the last decades have not brought any success (but at least some helpful insights).

The *simplest* conceivable solution to Problem 1 would be that a consistent QFT of gravity can be established under the *asymptotic-safety scenario*. The *simplest* conceivable solution to problem 5c would be that dark matter consists of a single particle species interacting with itself but not with the Standard Model particles (except via gravity), which can be described within the framework of a very simple QFT. In these cases, a new theoretical framework would not even be needed, everything would fit within the framework of QFT. In accordance with Occam's razor, one could consider these two approaches as working hypotheses for the time being.

Further Open Questions

But there are other unanswered questions that did not find their way into Smolin's list:

- How big is the universe? In particular, is it infinite? This is again one of those questions that, as mentioned before, we will almost certainly never find an answer to – because of the cosmological horizon.
- Why was the early universe so extremely uniform at the time when the background radiation was emitted, i.e. about 400,000 years after the Big Bang, across regions that were not yet in causal contact with each other at that time, because there was not enough time to send any signals back and forth between them, let alone to set any balancing processes in motion that could have ironed out the non-uniformities? Various answers are conceivable. The uniformity could be inherent in the initial conditions of the universe. Or it could have been dynamically generated by an effect of quantum gravity, which in a tiny fraction of a second was the competent theory at the beginning of the universe.

Interestingly, almost the entire community of physicists prefers a different explanation, namely a phase of **inflation** in which the universe exponentially inflated by a huge factor in a fraction of a second, "blowing away" all non-uniformities. Now, in turn, to explain how inflation occurred, a new field with some rather implausible properties had to be introduced. In my very personal opinion, the inflation hypothesis ranks somewhere between "highly speculative" and "completely far-fetched". I'm relatively lonely with this opinion, though, because most physicists already treat inflation like a fact. This becomes problematic when they discard plausible solutions to other problems because of inflation, as we will see in a moment.

I have never quite been able to understand the uncritical acceptance of the inflation hypothesis. Probably it plays a role that inflation also "saves" other theories in which many physicists would like to believe. I had already written that unified theories often produce unwanted relics that have to be "hidden" again by some other mechanism to explain why we do not observe them. One such relic, whose abundant existence the GUTs predict, is the so-called magnetic monopoles (you don't need to know what that is), but unfortunately they are undetectable in our universe. That's where inflation comes in handy, acting like a big leaf blower: The monopoles were all blown away by it, thus saving the GUTs. In the meantime, however, it doesn't look very good for the simple variants of the GUTs, also because of the undetectable proton decay. In the end, the rescue by inflation was possibly in vain.

No matter which explanation for the uniformity of the universe is the correct one, we will probably never know for sure because this early phase of the universe is completely inaccessible to our observations.

- Why is there so much matter left after the mutual annihilation of matter and antimatter in the early universe, but almost no antimatter? How does this asymmetry come about? The simplest explanation would be that this preponderance of matter was there from the beginning; the initial state of the universe just contained slightly more matter than antimatter. But because a large majority of physicists believe in the inflation hypothesis, this simple possibility is rejected: Inflation would have acted as a leaf blower here too and simply blown away the asymmetry. So the imbalance would have to have arisen after inflation. Unfortunately, the mechanisms that would be necessary for this are very complicated and require conditions that the Standard Model of particle physics does not provide. The least implausible scenario, so-called leptogenesis, presupposes certain properties of the neutrino, which at least can be tested in the not too distant future. Such a test will either rule out the possibility of leptogenesis or not, but it will definitely not be able to prove it, because several other things will have to come together to make it possible. In any case, it would be simpler to abandon the inflation hypothesis! As with all the questions discussed so far, chances are we will never know.

- What exactly are the properties of the neutrino? We already know quite a bit about this very light, elusive particle, but there are still some details we don't know. These include the exact mass (three different masses, to be precise, because it comes in three varieties), the knowledge of which would also help us to further constrain some cosmological parameters. Besides the mass, there are some other things to be clarified, but explaining them here would take too much space. These other things also determine whether leptogenesis is a possible scenario for explaining the matter/antimatter asymmetry, so they are definitely of cosmic importance. The neutrino details may not be among the most spectacular open questions in physics. But they have a very decisive advantage: it is realistic to clarify them in the foreseeable future! The experiments needed to do this are already being prepared (one of them is the KATRIN experiment in Karlsruhe).

- How can QFT be better understood? As already mentioned, we have well-functioning tools in the form of perturbation calculus (Feynman diagrams), which provide us with extremely accurate predictions for scattering and decay processes. But for several aspects of the theory, a deeper mathematical understanding is still missing. Among other things, it is very difficult in its present form to describe bound states, such as the hydrogen atom (bonding of an electron to a proton) – not to mention the larger atoms. The interaction of the quarks in the formation of the proton is also not yet understood, even though, as mentioned in Chap. 5, the numerical value for the proton mass could at least be reproduced correctly with supercomputers. A prize of one million dollars is offered for a solution of the problem that satisfies even the high demands of mathematicians; it is one of the seven so-called millennium problems.

This task seems to me to be the most important, most rewarding (and I don't just mean the million dollars) and most promising of all within theoretical physics. In the hunt for the theory of everything, it has unfortunately been somewhat marginalized. Yet there are many questions in physics that would benefit from it. We are dealing here with a quantum theory in which the structure of the state space is not yet well understood. With QM the meaning, i.e. the philosophical aspect, is unclear, here with QFT now additionally the mathematical one. Unfortunately, some of the speculative theories on the hunt for the theory of everything are even less clearly defined, so from this perspective QFT already looks like a haven of rigor and well-definedness. But it is not really so. A Pierre Duhem (Chap. 6) would never let QFT pass as a theory, at most as an intermediate step on the way to a theory.

8.3 The Crisis of Physics

Suppose you live in a true utopia: all diseases are cured, there is no war and no crime, social justice reigns everywhere, people live in abundance and all live to be 100 years old. All the knowledge in the world is available to everyone, everyone can exercise their creativity and work on the realization of their ideas – a land of bliss. In this utopia, humanity's problems all seem solved, except for one: people are aging and dying, still. And because it's the *last* problem, everyone is working on it; every conceivable approach that will bring them closer to immortality is being tried. It seems doable, people are optimistic. But then things go nowhere. The last problem turns out to be more and more unsolvable, not in principle, but from practical experience, from estimates of what would be technically necessary according to the knowledge gained versus what is technically feasible. The number of malcontents increases with each decade. Some shout slogans of perseverance. A certain disorientation arises. Little groups are formed who advocate their approaches to immortality more dogmatically and with propaganda than with scientific methods; or who bend the numerical values in their estimates to such an extent that they appear to be just within technically feasible reach. The land of bliss has become an unhappy land, and to some extent a dishonest one.

Fundamental physics is in a similar situation. The laborious process that Duhem has characterized (Chap. 6) has led to such successes that we now live in a physical utopia in which almost all fundamental problems have been solved, at least within the range that is experimentally accessible to us. The remaining problems are hard up against the limits of our possibilities or already beyond them. As a result, progress in physics has slowed considerably over the last 40 years. Many, especially in theoretical physics, speak of a crisis.

It has been apparent for some time that we are reaching our limits. Until about the Second World War, theory, experiment and application in physics always went hand in hand. From the steam engine to the light bulb to the atomic bomb, technology was always based on the *latest* discoveries in physics. Industry, general prosperity and the military depended to some extent on the natural sciences, and in particular on basic research in physics, also had an effect on them; practical problems inspired research. Or, in the more unpleasant version: basic research was demanded and promoted for military purposes. The interplay of theory and experiment also worked through and through: Experimental results had to be processed into theories, and the theories in turn had to be tested by new experiments. The new experiments showed some unexpected results, so the theories had to be refined again, and so on.

Nuclear physics is the last example in physics where theory, experiment and application interacted directly. After the Second World War, basic research in physics mainly meant particle physics and, somewhat later, cosmology. These fields were no longer "useful" in the sense of practical applications. You can't build a bomb out of antimatter, neutrinos, or dark matter (fortunately), or anything else that people enjoy. In exchange, they helped us understand what the matter we are made of is, and how it came to be. We have to keep in mind that all the great advances in technology we see around us are *engineering* feats built on physical principles that were known before World War II.

The lack of usefulness of the new results is already related to the fact that we are approaching our limits with basic research. All the new particles are not "useful" because they either decay into other particles in a fraction of a second or because they can hardly be "grasped", since they interact only extremely weakly with us, i.e. with the matter of which we and our environment consist. That is why they were discovered so late. They complete our theories, improve our understanding of the interrelationships on a small scale, but otherwise nothing can be done with them.

Until the 1970s, theory and experiment continued to go hand in hand, only the applications were out of the picture. That changed (at least in particle physics, not so much in cosmology) after a self-contained theory was found that successfully covered the whole particle zoo: the Standard Model of particle physics. This theory is so good that even better and better, larger experiments have not been able to produce results that contradict it, thus giving new direction to the work of the theorists. But that, if we follow Duhem, is precisely the purpose of theorizing, namely to advance "natural classification" by grouping experimental data into theories, giving them a rigorous, mathematical order. But what if all the experimental data are already summarized in the standard model? Then theory is left to its own devices, driven forward only by its own why-questions that want to trace theories back to even deeper theories. So while experimental particle physics was cleaning up after the Standard Model, confirming the existence of all the particles it predicted (the last being the Higgs particle in 2012), theoretical physics was stewing in its own juices, drifting further and further into speculation. Theory and experiment had become separated.

Time and again, *attempts* have been *made to* establish a link between theory and experiment, for example in the case of one of the most popular speculations, the so-called *supersymmetry*. This predicts the existence of further particles, and if the theory is correctly constructed according to the modular principle, these particles could have been such that they could have been found at the particle accelerator LHC. Unfortunately, nothing came of it, the new particles did not want to show up. It was hoped that these particles could also explain dark matter. But in the end it doesn't look like it, things don't fit together. Building theories according to the modular principle, without experimental evidence, has one big disadvantage: There are too many possibilities, and it is difficult to say when it is better to give up and throw a theory in the garbage. The next building block could always lead to success.

After all, cosmology brought the realization that 95% of the energy of the universe is in the form of dark energy and dark matter. However, these two things have only a very indirect influence on visible events, and this influence can be expressed in a few parameters. Dark energy is already compatible with existing theories anyway, and we can't really get at dark matter; we have to live with the few indirect clues we have. At least, some theoretical physics could be done with them, even if only to a small extent; some possibilities could be ruled out. But it remains to be seen whether the question can be resolved in a positive sense. It also remains to be seen whether an answer will reveal interesting physics or just a particle with very boring properties.

Side Effects

The stewing of theoretical physics in its own juice is not healthy and has caused unpleasant side effects. These include a certain dogmatism that binds some researchers to their speculations, sometimes assuming an almost confessional character and having more to do with faith than with natural science. Legendary and right to the point is the scene in the series *The Big Bang Theory* in which Leonard Hofstadter and Leslie Winkle discover that he is a follower of string theory, she a follower of loop quantum gravity (another speculative approach). *"How will we raise our children?"* they ask each other, and prefer to part ways; the denominational divide is just too wide.

This is not (or only in moderation) the fault of the theoretical physicists. The fact that they are stewing in their own juices is due to the great success of the Standard Model, that is, because they have done such great work. It is not easy to avoid these side effects with this starting position and the constant pressure that exists in research, considering also the social and psychological effects that we cannot escape. However, it is necessary to identify them and eventually correct them.

However, one must also emphasize that there are still very healthy, fruitful areas in theoretical physics. It is just that these have not been the focus of attention in recent decades. These are the areas that have not pounced on the "really big questions", but rather have gone a little more modestly into the already established theories, sounding out their consequences and connections, and thus in some cases even bringing new states of matter to light. For example, phase transitions between different states of matter have been better understood, matter near absolute zero temperature has been described more precisely, and deep connections between quantum theory and classical statistical mechanics have been uncovered. All of this is, of course, less spectacular than getting to the bottom of the Big Bang or grand unification, but it has a hand and a foot. One of the most popular areas, which has also made the most progress and attracted the most attention, is quantum information theory. In it, entangled states are used to transmit information that can be encoded better than in classical computers and also allow information processing in a new way (keyword *quantum computing*). This is "only" an application of an already known theory, QM, so it is not really fundamental physics, but there is a healthy exchange between theory, experiment and application!

Another side effect of the difficult situation in fundamental physics is the presentation in the media. Since (1) the theories have become so complicated that an outsider, even with a background in physics, can hardly get an idea of them, (2) there are so many convoluted assumptions and interdependencies in and between the various speculations that are difficult to see through, and (3) physics is dependent on research funding and needs to present its prospects in as positive a light as possible, it is almost a given that **propaganda will** ensue. Supporters of diverse speculations write books, where they present their theories in shining light, their beauty and elegance, the prospect of soon being able to explain the whole universe, on the other hand, however, the many remaining ambiguities and contradictions fall under the table or are at least strongly deemphasized.

On the experimental side, any measurement that seems to point to physics beyond the Standard Model is immediately leaked to the press with great fanfare. "Particles registered at faster-than-light speeds!" it was once said. Or, "The orbit of the Pioneer space probe shows a deviation from the law of gravity!" In all cases, the whole thing turned out to be a measurement error or the result of an unconsidered effect.

One case in March 2014 was particularly brazen. Some researchers claimed to have detected traces of gravitational waves in the background radiation. The news immediately went around the globe and was further hyped up. They had thus found proof of the theory of inflation! Alan Guth, one of the inventors of inflation, was already considered a Nobel Prize candidate. The enthusiasm evoked almost embarrassing comments: "This could be one of the greatest discoveries in the history of science!" How is an unsophisticated reader or viewer supposed to be able to properly classify something like this? A little more caution on the part of the press would certainly be appropriate.

The whole thing turned out a few days later to be a measurement error caused by dust. Even if the signal had been real, this would not have proved the inflation theory by far. First of all, gravitational waves are predicted by GR, not by inflation (so the actual measurement of gravitational waves one year later was correctly interpreted as the last missing confirmation of GR). Inflation indeed predicts that traces of gravitational waves should be found in the background radiation. The converse, that the existence of gravitational waves at this early stage of the universe is necessarily due to inflation, is far from valid. The inflation hypothesis would thus have survived only a falsification attempt, as Karl Popper would express himself. But with such a daring hypothesis, one would surely demand more evidence than that. And even if inflation had been "proven", it would still not be "one of the greatest discoveries in the history of science", but merely a new milestone in understanding the history of the early universe. So we have three interlocking gross exaggerations in this one report!

Another, not quite so drastic, example of the excessiveness of representation can be seen in the discovery of the Higgs particle. Of course, the operators of the LHC and physicists from all over the world were rightly delighted that this last building block of the Standard Model had been found, the only major success of the LHC. The success was savoured in the media. Several books were written about it. The "God particle" had been discovered, the media said. God particle! What an unspeakable name! What nonsensical associations it awakens! The name probably goes back to the fact that the Nobel Prize winner Leon Lederman wanted to title his book about this particle *The* Goddamn Particle. The publisher then put pressure on the author to rename it *The God Particle*, saying it would sell better. Equally ridiculous is the name "Big Bang Machine" for the LHC, which was heard more than once. The LHC does not simulate the Big Bang, it has nothing to do with the Big Bang at all, it simply shoots particles at high energies at each other and sees what comes out. Such pithy terms are meant to arouse people's interest, but they lead to false associations and misinterpretations, of which there are already so many, given the complexity and mathematical nature of the theories.

The crisis of fundamental physics is a consequence of its own success and the experimental inaccessibility – at least according to the circumstantial evidence – of the remaining big open questions. The ideological and media side effects are symptoms and consequences of this crisis, not its causes. Nevertheless, one should consider how to steer theoretical physics as a research discipline back onto healthier tracks. In my view, a little more modesty and recognition of one's own limitations would be the most important first step; in addition, more reflection on the deeper connections and less *model building* according to the building block principle, as well as a more precise, less spectacular presentation to the media and in the literature. This book is intended to contribute to that. By no means am I suggesting that we should stop working on the big questions. There is indeed hope that we will succeed in making a great breakthrough after all. On the other hand, we have to live with the *possibility* that we will never find an answer to many of the questions.

Books on the Crisis

Several other books have been published on the crisis in fundamental physics. I would like to compare their theses briefly with my own:

- *The end of science* by John Horgan (1996): Horgan has taken a lot of flak for his thesis that the age of great scientific discoveries must eventually come to an end, although this thesis is clearly correct. He explicitly refers to the findings of basic research, not to technical applications. There are only finitely many different phenomena to describe in the part of the universe we can observe. The explanation of some of these phenomena may be forever beyond our practical ability, the others will all be eventually understood. The whole methodology of science is geared to consensus and unification, to convergence on a single overall picture. It is therefore impossible to keep throwing everything over and starting again. Horgan sees signs that the progress of knowledge has already slowed down considerably, that we are thus gradually approaching this end of natural science. For the recent decades of theoretical physics he uses the term **ironic science**, meaning a science that is actually no longer a science because it is purely speculative, purely a matter of interpretation, and unverifiable. On the experimental side, although the abundance of data is increasing enormously, the details and decimal places are being filled in more and more, but fundamental insights have been gained less and less; hardly anything has changed about the world view as a whole.

It is in itself a good sign that we are gradually converging on a consistent overall picture – within our limits; what lies outside must remain speculative, "ironic". The time of the great discoverers *must, after* all, be over at some point (except in mathematics, which is infinite). How far it is in biology with it, I will not judge here, there the variety is much greater than in physics; I am concerned here only with the latter. And here I see a lot of evidence that Horgan is right. I may be wrong, of course. There may be things we simply haven't thought of yet, clever precision experiments with which we can get to the bottom of quantum gravity after all, or a variety of new phenomena just at the next energy scale to be tackled with the next particle accelerator, or a dramatic deviation from the Standard Model unexpectedly turning up somewhere else after all. All of this is possible, and it's also important to keep looking for it. It's more of a toss-up: how likely do we think such an event is, given that nothing like it has been found in the last 40 years, despite tens of thousands of researchers involved in the search in many different ways? And if the search remains unsuccessful for another 40 years, how many researchers will still be found to take part in the arduous and, if unsuccessful, unsatisfactory work? How much money will still be made available to such projects?

- *Not Even Wrong* by Peter Woit (2006): In this book Woit describes how one of the speculative approaches, string theory, has gone off the rails. In particular, he lets loose about the huge discrepancy that exists between the self-assured, arrogant demeanor of many string theorists and the actual accomplishments of the theory. After all, he says, it's not even a real theory, just a series of conjectures that are not even mathematically understood and whose consistency has not been proven. Moreover, none of it can be experimentally verified. Therefore, it is "not even wrong", because in order to be wrong, one would first have to make a clearly defined theory out of it and show how it could be tested. String theory sees itself as *"the only game in town"* and is perceived as such by many outsiders. This is why young researchers are under pressure to work on it, even though other approaches might be more promising.

From my point of view, string theory is a theoretically highly interesting approach both to quantum gravity and to the unification of forces, based on generalizations of the methods of QFT. Since, as I said, even in QFT much of the fundamentals are not yet conceptually understood, it is not surprising that these ambiguities are inherited and amplified in string theory as well. It is indeed an extremely complicated theory. It basically plays out on the Planck scale and thus, *for practical reasons,* defies experimental verifiability. But this is neither the fault of the string theorists nor a problem of string theory itself, because the Planck scale is by all theoretical evidence the scale of quantum gravity. Other approaches to it are likely to suffer from the same problem. *If* we could do experiments at the Planck scale, then there would also be predictions of string theory that we could check, for example that spacetime is ten-dimensional.

Yes, you can ask yourself whether this is still sound science. But that's just the way things are in physics. It would have been nice, of course, if the theory at least predicted an unambiguous *effective theory* at the scales we have access to. But it turns out that there are a huge number of ways in which the theory can be "broken down" from the Planck scale. Again, this is not the fault of the string theorists. They follow the approach that seems most plausible to them (and for reasonable reasons, to the best of their knowledge), and see how far it gets them. Along the way they have made valuable contributions to mathematics and also produced a better understanding of certain quantum field theories. To condemn these efforts as nonsensical and a waste of time seems rather arrogant itself.

The side effects (exaggerated self-assurance, dogmatism, disdain for other approaches, pressure on young researchers) are certainly present. Overall, however, it seems to me to be primarily an American problem. I have talked to quite a few string theorists in Europe, and none of them have lived up to the arrogant stereotype Woit paints. In Europe, string theory has never been considered *"the only game in town"* to the same extent; the relationship between the various approaches is much more balanced.

Modesty is indeed appropriate (and is generally a healthy attitude in the natural sciences, and elsewhere) when dealing with such speculative approaches. However, I think it is fundamentally wrong to blame the crisis in physics on the behaviour of string theorists.

- *The Trouble with Physics* by Lee Smolin (2006): This book has already been mentioned several times; in it Smolin first describes the understanding of values in the natural sciences and what he sees as the major open questions in physics. The main part of the book, however, is about the crisis of physics. Initially, Smolin also discusses the problems of string theory. Overall, however, he aims at a much more general issue. In his view, a new scientific revolution is needed to resolve the open questions, of similar magnitude to the Copernican turn, Newton's laying of the foundations of physics, or QM. But the current scientific establishment is not designed for revolutionary research, he says. Young researchers are forced to publish as much as possible as quickly as possible and to follow the mainstream, otherwise they run the risk of not getting a permanent position. For him, this is the reason for the crisis, the lack of progress in recent decades. If we were better able to promote "revolutionary" research, we would be able to make a big impact.

That seems to me to be a misreading of the situation. The lack of progress is due to the fact that the big open questions take place near the Planck scale, which is not accessible to us, or near the Big Bang, which is also not accessible to us, forcing us to make speculations that cannot be verified. I do not see the slightest evidence for the need for a "scientific revolution". Rather, I have argued that the remaining big questions, as far as they are scientific questions at all, may have fairly simple answers that fit within the framework of currently known theories. And even if the answers are a bit more complicated, they might take a form, for example, that follows from string theory, which is also not "revolutionary" enough to Smolin's taste. The main problem, after all, is the lack of verifiability, the lack of experimental evidence, and that is due to the nature of things (or their scale) and our *practical limitations* in experimenting, not primarily to the nature of our research. That being said, I agree with him in that the building-block approach currently being used to construct many models is not conducive to progress.

- *Vom Urknall zum Durchknall* (in English available as *Bankrupting Physics*) by Alexander Unzicker (2010): This is a pretty blatant polemic in which Unzicker lashes out at everything that physics has produced in the last 50 years. It reads quite entertainingly, but overshoots the mark by far. Even the quark model and all the findings of cosmology are called into doubt, which have meanwhile been confirmed by very different observations with many mutual checks and balances. Certainly one can doubt one or the other experiment, one or the other observation, but not the consistent overall picture that results from their often redundant combination. So the claim that all of physics took a completely wrong turn 50 years ago doesn't seem very well founded, rather insufficiently informed. The approaches

Unzicker proposes for the "right" way try to turn back time 50 years and seem naive.

- *Lost in Math* by Sabine Hossenfelder (2018): This book argues that the lack of progress in recent decades was due to theoretical physicists asking the wrong questions and pursuing the wrong approaches, for one particular reason: they were blinded by their claim to aesthetics and mathematical elegance. This gave rise to the claim that the answer to all questions must be "beautiful", and every "ugly" aspect of a theory, such as the multitude of parameters of the Standard Model, was regarded as a problem yet to be solved. This pushed real problems into the background. We also had to accept ugly solutions without bias. This is why the book is also called *Das hässliche Universum (The Ugly Universe)* in German, in allusion to and as a counterpart to Brian Greene's *Das elegante Universum (The Elegant Universe)*.

This thesis has something to it; many problem definitions and approaches to solutions seem somewhat ideologically coloured. I had criticized some of Smolin's big questions accordingly. The claim, the wishful thinking we have of a theory, may or may not be true. On the other hand, elegance is also a useful criterion in physical theories. Occam's razor requires the *most efficient* summary of phenomena. Now, efficiency and mathematical elegance are almost synonymous in such highly mathematical constructs as the theories of fundamental physics. The search for a theory with fewer parameters than the Standard Model is at once a demand for elegance and for efficiency. In the highly speculative areas in which theoretical physics must currently seek its progress, mathematical elegance may even be one of the few criteria by which one can be guided at all.

In no way can I agree with the author that this is the fundamental problem with the current crisis. Even in the questions she considers to be the real ones (which are essentially the same as my own selection), even in the ugliest approaches that have been taken to answering them (and which have by no means been shunned by all), there are no great successes to report. The real problem remains the lack of verifiability due to the inaccessibility of the relevant scales.

Thus, I only agree with John Horgan in his analysis of the causes of the current situation in physics. We are hard up against the limits of what is feasible in physics to the best of human ability. There is a realm beyond that which will remain inaccessible to us. There is some evidence that many of the questions in physics that are considered important today already have their answers beyond that limit. It is great that nevertheless so many researchers have not given up courage and are trying. But for how much longer?

8.4   The Multiverse

Conditions on planet Earth are decidedly conducive to life. There is water in abundance, enough carbon for the formation of complex chemical compounds, the temperature is not too hot and not too cold, the composition and pressure of the atmosphere are just what we need, a magnetic field shields us from much of the cosmic radiation, and even the moon stabilizes the Earth system in many ways. That's just a small part of the list of what factors on our Earth are just perfectly aligned for the emergence of life. Of course, one might think that a god created this wonderful system to provide us with a home. But today we know that there are already billions of planets in our Milky Way, not to mention the rest of the universe. On these planets, a wide variety of conditions prevail in a wide variety of combinations. Among all these planets, *we* naturally live on one that is well suited for the emergence of life, otherwise we would not be here after all. This is the so-called **weak anthropic principle:** a wide range of possibilities is realized, and we necessarily find ourselves on one that admits of intelligent life, no matter how improbable that constellation may be in terms of the total. That is, our planet was not made especially for us, but is simply a chance product in the vast variety of planets that exist in the universe. And surely there are intelligent life forms on many other planets that come to similar conclusions.

In the case of planets, the weak anthropic principle is clearly a *scientific* statement. This is because we can *observe* other planets, primarily those in our solar system, of course, but now indirectly also those in other star systems. We can compare their properties with each other, can roughly estimate distributions and frequencies on the basis of astronomical and physical theories, and can thus quantify (albeit still rather inaccurately) *how* exceptional the Earth is in the total set of planets.

But what if we could not see any other planets (i.e. if the sun had only one planet, namely the earth, and also no exoplanets were known)? If we could not even be sure that other planets existed at all, i.e., if this were a pure hypothesis? What would then be the character of the weak anthropic principle? The question is quite subtle. If we assume that it may be possible to observe other planets in the future, just not yet, then it remains a scientific hypothesis. However, if we had no evidence for the existence of other planets, and also knew that we would *never* be able to observe them if they did exist, then we would probably dismiss the principle as *metaphysical* from the natural sciences. We could still consider it *plausible* and contrast it with the equally metaphysical hypothesis of a special divine creation of our planet, but it would then no longer be a scientific question.

However, when the state of knowledge is somewhere in between, it gets complicated. For example, if we had astronomical theories that favored the existence of other planets, theories that were successfully confirmed based on *other* predictions, but we still knew that the planets were one of those predictions that we could *never* confirm, then the principle could just about pass as scientific, precisely because the existence follows from *confirmed* theories (this is the main argument Max Tegmark makes in various discussions for the multiverse). If, however, the theories are not confirmed or the interpretation of the theories is disputed, then we are in a borderline area.

This is exactly the discussion about the possible existence of other universes that is currently being held. The core of the question is really whether this is still natural science. We find that the parameters in the theories that describe our universe seem to be tuned just right for the emergence of life. This shifts the whole discussion we just had about planets to the universe as a whole. The weak anthropic principle then says: there are many universes (all of which together then form the **multiverse**), with a wide range of natural laws, and we naturally find ourselves in one in which the emergence of intelligent life is possible.

To do this, we should first clarify how the terms universe and multiverse are used. For Max Tegmark, one of the most important proponents of the multiverse idea, *another* universe is a world (however defined) with which we will never come into contact, and which we will therefore in particular never be able to observe. He gives several examples of this. Thus, for him, the many worlds of the many-worlds interpretation of QM (of which Tegmark is a follower) form a multiverse because these different worlds are no longer in causal contact with each other or ever will be again. Even regions of the universe that are so far apart that they can never communicate with each other already form a multiverse for him. One can already see here that the terms universe and multiverse are not so clearly demarcated. The many worlds of QM together form the state of a single universe, they are only different *branches of* this one state. The distant regions of the universe are part of a common spacetime continuum. There is really no need to introduce a new term here, that of multiverse. It is a *universe* in which there are just parts that are not in contact with each other.

Another stage of the multiverse is reached when the distant regions follow different *effective* laws of nature. That is, the same fundamental unified theory (the same *theory of everything*) underlies all these regions. But since such theories are often defined at very high energy scales, as I said, and can be broken down to lower energies in a great many different ways, it may be that different regions of spacetime realize different of these possibilities, and thus *effectively* have different laws of nature (unless you do experiments at the scale of the unified theory, in which case the laws of nature are the same again). This is the level that is relevant to the weak anthropic principle and is what most refer to when talking about the multiverse. In this context, the term *universe* is used to refer to a region of spacetime with certain uniform physical properties, and the term *multiverse is* used to refer to the spacetime continuum as a whole.

Is this still natural science? Proponents like Tegmark argue: If we could find a unified theory and confirm it experimentally in some way, and that theory now has the property of allowing for very many different possible effective theories at low energies (as is the case with string theory, for example), then we would have a natural science handle on this kind of multiverse. One would then still have to clarify how it came about that different possibilities were realized in different regions, but that would also be a question that could be investigated in the context of the unified theory.

The "if" that stands at the beginning of this argument is, of course, a very big one. At the moment, we are far from being able to experimentally confirm a unified theory. In this respect, the multiverse idea is also pure speculation at the moment. But it is indeed possible in principle that the topic can be treated ("only just") scientifically, if one defines the multiverse as above as the entire *one* spacetime continuum.

But the multiverse idea has also been pushed even further, beyond the one spacetime continuum. Tegmark goes so far as to suggest that *every* mathematical structure forms a universe that *exists in* some Platonic sense. The totality of all mathematical structures thus forms a gigantic multiverse. Our universe is one of these infinitely many structures. This is, of course, a purely metaphysical speculation. However, it has some interesting philosophical implications, so we will return to it in the next chapter.

9

Things and Facts

9.1 Facts

"The world is everything that is the case. It is the totality of facts, not of things", says Wittgenstein (1963) in the already quoted *Tractatus*. But what is a "fact" after all? In mathematics, statements have a definite truth value (true or false), definitions are unambiguous, proofs leave no room for doubt. It seems to be similar in criminology: either Lee Harvey Oswald is Kennedy's killer, or he is not, even if we don't know the answer. Crime novels like those by Agatha Christie or Sir Arthur Conan Doyle suggest that pure logic rules here, coupled with empirical findings; circumstantial evidence pieced together to form proof. The whole of natural science seems in this sense to be one big detective series, on the constant hunt for facts.

But already in the case of crime it is often not so clear. Did the perpetrator act with intent or not? Was he sane at the time of the crime? The legal code demands a clear classification, but there are borderline areas where the answer seems to lie somewhere between yes and no. The more subjectivity comes into play, the more ambiguous the facts. That's why natural science strives for perfect objectivity.

However, in physics it is very difficult with the facts, despite objectivity. There are many reasons for this. All of them have to do with the fact that there is much more **context** attached to physical statements than to statements referring to everyday situations, such as murder and manslaughter, and the further down the hierarchy of theories you go, the more context there is. We want to discuss here the different kinds of context that occur and what this says about physical facts.

© Springer-Verlag GmbH Germany, part of Springer Nature 2022
J.-M. Schwindt, *Universe Without Things*,
https://doi.org/10.1007/978-3-662-65426-2_9

First Complication

The first complication is the **experiment,** which stands between the physical statement and the physicist. During observations in everyday life, we perceive facts directly through the senses. We perceive colours, shapes and sounds and directly recognise objects, people, melodies and many connections between them. In a physical experiment, the relationships are more complicated. As already described in Chap. 6, the result of an experiment is not simply the record of the experimenter's sensory perceptions. Instead, these perceptions, e.g. the reading of a measuring device, are *interpreted* as statements about certain physical quantities that cannot be perceived directly by the senses, e.g. the strength of a magnetic field. The magnetic field is itself already something conceptual, abstract, the result of physical theorizing. The experimenter and also the reader of his report must already have the associated concepts and connections in mind in order to understand what the measuring instrument actually does, i.e. how the statement about the magnetic field that the experimenter writes down is derived from it.

The magnetic field is still a relatively simple example. But if you ask a particle physicist how exactly, according to which logic the data output of the detectors at the LHC can be understood as a proof of the Higgs particle, the answer will be extremely complicated. The result of an experiment, the statement it makes, can only be understood in the **context of a theory** (or, in complicated cases, even several theories). The theory, in turn, is based on a chain of other experiments, as well as on a series of conceptualizations and the recognition of connections between those concepts. The concepts are abstract, i.e. they can be illustrated to a certain extent, e.g. by means of the magnetic "field lines" in the case of the magnetic field, but they are outside the world of our sensory perceptions and involve a certain amount of effort for the human mind, and require some practice.

Second Complication

The second complication is that both measurements and theories are **approximations**. Any measuring device has a limited resolving power and may be subject to certain perturbations. Each experimenter has to estimate the inaccuracies associated with a measurement. This estimation is an essential part of an experimental protocol. Therefore, any statement about a physical quantity must always be implicitly understood to have some level of imprecision. If my identity card says that I am 186 cm tall, this does not mean that I am exactly

186.0 cm tall, but between 185.5 and 186.5 cm. Numerical values are always to be understood in the sense of roundings, not as exact values, and are therefore handled differently than in mathematics (Chap. 6).

Theories refer to such measured variables that are subject to inaccuracies and are thus themselves only to be understood as approximations. This is also how the hierarchy of theories comes about. A theory can describe a set of experimental results well within the limits of its accuracy; however, if the accuracy of the experiments is increased by improved methods, the theory may no longer be sufficient and must be replaced by a better, more accurate theory. Moreover, how well a theory works as an approximation depends on certain features of the physical situation. When it comes to macroscopic objects whose velocities are much smaller than the speed of light, Newtonian mechanics is a good approximation. The more the velocities approach the speed of light, the worse it becomes. In that case, it is better to use SR. This is a good approximation at high speeds, provided a few other conditions are met, for example: distances are small compared to the size of the visible universe, and gravity does not act too strongly on the objects, otherwise you would need GR, as an even better approximation. Among the planets of the solar system, Mercury is subject to the strongest gravity from the Sun. Its orbit was so accurately measured more than 100 years ago that Newton's theory of gravity was no longer accurate enough to predict it correctly.

When I say that every theory is to be understood as an approximation, one may ask: approximation to what? To "reality"? But this is a problematic concept. Theories put experimental findings into context, but to what extent they represent a "reality" is a highly philosophical question. Perhaps: approximations to findings as *ideal* experiments would find them, i.e. those with vanishing inaccuracies? This is also very problematic. On the one hand, such experiments are practically impossible, and on the other hand, even some theories already have in-principle inaccuracies built into them, as in the uncertainty relation of QM. This relation implies that infinitely accurate measurements require interactions with infinite energy, which would collapse the whole experiment into a black hole. We should therefore be a little more modest and understand the approximation gradually: Every theory is an approximation to the next "better", i.e. more accurate, theory.

What does it mean that a theory T1 is an approximation to another theory T2? The two theories may use completely different conceptual systems, so it may not be possible to simply compare a statement of one theory with another. The link is again the experiment: in order for the theories to be comparable, they must be able to be applied to the same experimental data, even though they may interpret them in terms of different concepts. One can then define

T1 to be an approximation to T2 if (1) in some cases both T1 and T2 describe or correctly predict relationships between data within the bounds of their accuracy, (2) in other cases only T2 does, and (3) never only T1 does. Finally, (4) there may be data about which T1 can say nothing at all, not even approximately, but (5) there are no data about which T1 can say anything but T2 cannot.

Let us go through this concretely with an example: When planetary orbits are measured by astronomers, this already involves a certain amount of theoretical knowledge, which we will summarize as T0. T0 mainly includes the optics necessary to understand how a telescope works. From observations over a period of years, the planetary orbits are inferred. These are interpreted as orbits around the sun, not around the earth, for example, following Copernicus' findings that the earth is not the center of the universe. I also count this as a component of T0.

These findings were already available to Kepler in the seventeenth century. In the measured orbits he found certain correlations, which he summarized to a theory T1, consisting of the three Kepler's laws. These laws describe the orbits according to their geometric properties and the time in which they are traversed.

A few decades later Newton found the law of gravity T2. This law traces the orbits of the planets back to a *force*, the gravitation. From this force, Kepler's laws for the orbits follow at first, but additionally the influences of the planets among each other are taken into account, which lead to small deviations from Kepler's laws, so that T2 is more accurate than T1. Depending on the accuracy of the measurement, T1 is sufficient to correctly reproduce the relationships between the data (case 1 in the definition above), or not (case 2). T1 therefore represents an approximation to T2. However, the law of gravitation is much more general than Kepler's laws; it also describes the falling of a stone on Earth, which Kepler's laws have nothing to say about (case 4).

Newton's law of gravity T2 is again an approximation to GR, which we label T3. This describes the orbit of Mercury more accurately than T2 (case 2), but for the outer planets the predictions of the two theories differ so little that the difference is below the accuracy of measurement (case 1). In GR, there is no longer a Newtonian force attracting the planets. Instead, they move along geodesics in a curved spacetime. The conceptual framework has changed a lot again. GR also describes situations to which Newton's theory has nothing to say, for example the accelerated expansion of the universe by dark energy (case 4).

We know that GR is not compatible with QM and can therefore devise experiments which are not currently feasible in practice but are feasible in *principle*, and of which we know that T3 (i.e. GR) will not be sufficient to correctly describe the relationships between the results of the experiment. A theory T4 will then be necessary, already nicknamed *quantum gravity*, to which T3 is an approximation.

One can of course wonder whether this chain of approximations will come to an end at some point, i.e. whether there will be a **theory of everything** into whose purview all experiments and observations that can in principle ever be carried out will fall, and which will correctly predict the results of all these experiments within the limits of their accuracy. In principle, both are conceivable: that the chain of approximations continues to infinity, or that it ends in a theory of everything. Regardless of this, in both cases it may be that we get stuck somewhere in the chain for *practical* reasons, because we are not able to perform the necessary experiments with the necessary accuracy (as described in Chap. 8).

Two things are clear, however:

1. Currently, we are not in possession of the theory of everything (if there is one) because we are not even in possession of a theory of quantum gravity.
2. If we have found the theory of everything, we shall never be able to *know* that it is the theory of everything, for we cannot know whether some phenomenon not yet observed by us contradicts it.

The theory of everything (if it exists and if we are in possession of it without knowing that it is the theory of everything) may well become a *confirmed* theory for us, which we take as sufficiently verified according to our common sense. However, we have only ever covered a certain area of the universe with our experiments and observations, only certain length, time and energy scales. Therefore, the theory can only be considered confirmed in this limited *range*. We do not know whether other phenomena exist beyond it that call for other theories. We can extend the range to some extent with future experiments. But the range we do *not* know will at any time be very large, possibly larger than that which we do know (Chap. 10).

A workable way to delineate the jurisdictional areas of each theory from one another uses the notion of **scale.** We know length, time, and energy scales, among others, taking advantage of the fact that length, time, and energy are concepts that appear in *all the* theories we have confirmed, so we can use them across boundaries. Better yet, length and time are closely related via the two theories of relativity, and energy in turn is closely related to time via QM. It so happens that QFT, which includes both QM and SR, has only *one* notion of scale, which can be understood as either length scale, time scale, or energy scale. (Where the latter is antiproportional to the other two: A large length or time scale means a small energy scale, and vice versa). It should be noted, however, that the three terms undergo certain changes when translated between different theories, that is, they have slightly different meanings. For example, mass is included in energy in relativistic theories, but not in non-relativistic theories. Spatial lengths and temporal distances in non-relativistic theories refer to an "absolute" space and an "absolute" time neatly separated from it. In SR and GR space and time are related, spatial and temporal distances depend respectively on the reference frame. In GR, moreover, distances must generally be calculated along curved lines in a curved spacetime. The notion of scale, however, refers to **orders of magnitude** and disregards such "subtle" differences. When used across theories, it is itself only a rough approximation.

Length scales are the easiest to talk about. Above 100 million light-years, for example, we are on cosmological length scales; here, GR rules above all, and everything can be characterized on the basis of the few cosmological parameters. Indeed, doing physics on a certain scale also means not taking into account non-uniformities that are much smaller than the scale under consideration (on cosmological scales, this would be the galaxies, for example). If we look at the shape of the Earth on the scale of 10,000 km, it appears as a uniform sphere. However, at the scale of 10 km, the irregularities, mountains and valleys, become visible. Our everyday life typically takes place on scales between centimeters and a few kilometers; these are the sizes of the objects we handle, the distances we travel. Here, the laws of classical physics mainly apply. Below nanometers (millionths of a millimeter) we enter the atomic scale. Here, the laws of QM govern. So different laws apply to different scales, different areas of physics are responsible.

In QFT, scales are usually expressed as energy scales, in units of electron volts (eV) (Sect. 8.1). What is typically meant here are the energies with which particles are shot at each other in order to test a theory. This also shows that there are scale dependencies at different levels. First, there are the continuous "running parameters" of QFT: parameters such as masses and coupling constants are functions of the scale parameter (Sect. 7.7). Second, there are so-called phase transitions, certain values of the scale parameter at which the behavior of the theory changes abruptly. For example, above the electroweak scale (about 100 GeV), the electromagnetic interaction and the weak nuclear force are united into a single force; below the scale, they differ. But even with such phase transitions, the fact that the underlying theoretical framework remains QFT does not change. Third, there are also scales where the entire theoretical framework changes. For example, if a QFT of gravity turns out to be impossible, then above the Planck scale (say 10^{19} GeV) the regime of quantum gravity is expected to set in, which is no longer a QFT. At the other end of the spectrum, in the 1 eV range, chemical reactions take place (i.e., these energies are typically exchanged or released per particle in chemical reactions). But these are much better described by QM than by QFT, so the theoretical framework of QFT is abandoned on this side as well. Not that QFT becomes *invalid* here, any more than quantum gravity becomes invalid below the Planck scale. There is merely a *simpler* theory which is quite sufficient as an approximation in this range.

The energy scales of QFT are directly linked to length scales, one being proportional to the reciprocal of the other (the larger the energy, the smaller the associated length). For example, the Planck scale corresponds to a length of about 10^{-33} cm, the *Planck length*. Thus, for the sake of uniformity and clarity, we can also speak of length scales in this area.

It may happen, by the way, that one day a new theory will appear in which the concept of "length" no longer appears. Then, at this point, we would also have to give up the concept of length scale, which has accompanied us through all theories until then and served as a distinguishing criterion.

To summarize: Theories are to be taken as approximations, as are the results of experiments. Moreover, each theory has a certain range of competence in which it is a *good* approximation. This range of competence can usually be characterized in terms of scales. Since physical statements are always in the context of a theory, they themselves represent approximations; in particular, their validity is bound to certain scales. With greater accuracy of the experiment to which a statement refers, or with deeper "zooming in" on the object it refers to, it may lose its validity.

Third Complication

The third complication is that physical quantities are relative. They are defined only in comparison to other quantities. If I say my shelf has a height of two meters, then not only must we take into account that this is an approximation, but we must first know what a meter is in the first place. The meter must be defined in some way. That my shelf has a height of two meters means that it is twice as high as what is defined as one meter, so it is a comparison with a different length.

So how is a meter defined now? That has changed over time. Originally it was once "the forty-millionth part of that circumference of the earth which touches Paris and the North Pole". The naming of Paris and the North Pole is necessary because the circumference of the Earth is not exactly the same everywhere, it has slight differences depending on which great circle you measure it along. Now, however, the surface of the earth is subject to certain small changes. The North Pole shifts a little from time to time, the continents drift back and forth, thus Paris also shifts, though very slowly. Let's assume that this shift causes the circumference of the earth, which meets the definition, to increase a little (only by a tiny fraction, of course, but that's not what matters here). Then it means that a meter has become slightly longer than before. Compared to this new meter, my shelf is then no longer two meters high, but slightly less. The shelf has shrunk a bit compared to the definition of the meter. But since the meter first defines what a length statement means, it is actually already wrong to say that the said circumference of the earth has increased. The meter *defines* if and how much a length changes. We act as if "length" has an absolute meaning, of which we can make a statement independently of all other factors. But length only ever has a meaning in comparison to another length. So if we say we define length by comparing it to said circumference of the earth, then *by definition* that circumference of the earth *cannot* grow. Rather, we would always have to assume that the rest of the world has shrunk. As a result, we would need new physical theories to explain why everything in the universe is shrinking. It would then turn out that these theories would be, first, very complicated, and second, that the shrinking of even the most distant stars is strangely related to continental drift on Earth. We would eventually realize that the strangeness of these theories was due to our unfavorable definition of the meter and change the definition.

Today, the meter is defined by the speed of light, as "the distance light travels in 1/299,792,458 of a second." The speed of light is assumed to be a natural constant that never changes. With the new definition of the meter, it cannot change *by definition*. If we should one day find that everything in the world has shrunk (as compared to the distance light travels in 1/299,792,458 of a second), then we have two choices: Either we look for a theory that explains why everything shrank without changing anything in the definition of the meter. Or we replace the definition of the meter with something else that cancels the shrinkage, and in exchange accept that the speed of light has changed, for which we then also need a new theory. Which of the two options we choose depends on which version ends up being more efficient, so in a sense it's Occam's razor. If both variants turn out to be equally complicated, then it is a *matter of taste* whether the speed of light has changed or the world has shrunk.

In reality, the matter is even more complicated, because the definition of the meter includes the unit of measurement "second", which in turn also needs a definition. The second is currently defined in terms of a quantum mechanical transition in the caesium atom. However, the physical properties of caesium would change if the atom shrinks as assumed, and this would also upset all time measurements. This change would also have to be taken into account. It turns out that via the laws of physics and the definitions they presuppose, pretty much everything is linked to everything else, and one can only assess all the changes in context, not simply the changes in length on their own.

Let's do another thought experiment on this. If overnight everything in the universe became twice as big as it was before (including all distances), would we even notice? The answer is: it depends on what else changes. The proportions stayed the same because everything grew at the same time. But what about gravity, for example? If the masses of the objects haven't changed, but really only the sizes, then it follows from Newton's law of gravity that the force of gravity on the surface of the Earth is only a quarter of its previous strength because the distance from the center of the Earth has doubled (the force of gravity is antiproportional to the square of the distance). So everything falls four times slower. However, since all scales have doubled at the same time, for example also the height of the tower from which I drop a stone, it seems even eight times slower to me. So in order for me not to notice any difference, there must be other things that change besides the lengths. For example, if the gravitational constant increased by a factor of 8, that would exactly cancel out the decrease in perceived gravity. Or the gravitational constant has stayed the same, but the mass of the Earth has increased by a factor of 8. That's plausible if the density stayed the same. Then a doubling of the radius just causes an eightfold increase in mass. In those two cases, you wouldn't notice any difference as far as gravity is concerned. However, one would have to make the same consideration with respect to all other laws of nature. In the end, one comes up with a combination of physical quantities and natural constants that would have to be changed *together* so that the doubling of lengths would not be noticeable. One could then justifiably claim that the universe after the change is the *same* as before.

It is remarkable that this is possible. It shows that the units of measurement and natural constants must be seen in a common context. One can then find **transformations** that **translate** one set of statements into another set of statements, so that the two sets, although they look completely different, describe the *same* physical situation. Take, for example, the expansion of the universe. Expansion means that all distances in the universe are increasing, but "smaller" objects, such as galaxies, which have "decoupled" from the expansion, retain their size. Can this situation be reinterpreted using tricks similar to those above to mean that distances are not increasing at all, but that, on the contrary, galaxies are shrinking? Indeed, it turns out that this is possible if one reinterprets a number of quantities and makes some natural constants time-dependent (Wetterich, 2013). The theory that emerges is not that much more complicated or implausible than the original one. The upshot is: whether one speaks of an expanding universe or a nonexpanding one in which things are shrinking is a *matter of taste* or *convention*. Both views are equivalent. That's how relative physical quantities are.

Fourth Complication

The fourth complication has to do with the third, is in a sense a generalization of it. It consists in the fact that the "physical core" of a theory – or of a statement made in the context of that theory – is clouded by many things that look to the uninitiated as if they belong to the physical core, although that is not the case. The "cloud" becomes thicker and denser the further down the hierarchy the theory lies.

Three components can be distinguished about this cloud: First, it contains things that are pure convention. These include coordinate systems. In order to describe a constellation in space (let's stay with classical mechanics or field theory for the moment), the physicist uses a coordinate system with which he "names" points in the form of combinations of numbers. Coordinates are pure convention. The physicist can use a Cartesian, spherical, cylindrical, or oblique-angled coordinate system, and he can place it in space as he pleases, i.e., choose the origin (where all coordinates are 0) as well as the orientation in space. Now, if he makes the statement that two objects collide at the point $(0,1,1)$, the three coordinates 0, 1, 1 belong to the cloud, not to the core. They spring from convention. To get to the core, one would have to translate them back into a physical characterization of the point using the definition of the chosen coordinate system, for example "one meter above the ground, one meter north of the measuring device".

The transition from one coordinate system to another is a *transformation* that *translates the* same physical situation into another convention, similar to a translation between two languages in which the same thing is to be expressed. Such transformations play an important role in physics, occurring in much more general contexts than just coordinate systems. We have already seen more complicated examples above: Changes in lengths that are translated into changes in something else by transforming various physical quantities, without anything having changed in the underlying physical situation.

The second component of the cloud is the dependence of physical quantities on the perspective of the observer. While this is closely related to coordinate transformations (change of reference frame), it is not quite the same thing. In classical mechanics, for example, velocities depend on perspective. Velocities are *relative* to a reference frame. It does not matter whether the reference frame is expressed in terms of a Cartesian, spherical, cylindrical, or oblique coordinate system. Reference system is therefore a less specific term than coordinate system. Instead, it expresses something physical: A reference frame always represents the perspective of *something* or *someone*. When we talk about velocities on Earth, we usually mean velocities relative to the ground, i.e. the velocity from the perspective of a potential observer standing somewhere on the ground. However, 2/3 of the earth's surface is covered with water, and there speeds are often relative to currents, as every swimmer knows who has ever been swept away by a current. A ship moves slower upstream than downstream, because its engines work relative to the current, not relative to the ground.

Forces are also dependent on perspective in classical mechanics. When you sit in a carousel, you experience a centrifugal force *outward*. To an outside observer, however, you are forced into a circular path by a centripetal force inward. However, one finds a class of "good" reference systems that move uniformly relative to each other, in which only very specific forces act, the *fundamental forces* of nature. This observation makes it possible to dismiss all other forces as *pseudo forces*, which only result from the fact that one has chosen an "unfavorable" reference frame, namely one that is accelerated relative to the "good" ones. The "good" reference frames are called *inertial frames*.

The crucial point now is that when descending the hierarchy of theories, things become perspective-dependent where this contradicts our intuition, because this intuition is built up from everyday experiences in which these differences do not occur. This becomes clear, for example, in SR: the time that lies between two given events suddenly depends on the frame of reference! Even relative velocities are now a matter of perspective (so velocities are now relative in a double sense). If Erwin and Otto are both flying away from Rita, each at half the speed of light but in opposite directions, then the relative speed between Otto and Rita is (1) from Otto's perspective 50% of the speed of light, (2) from Rita's perspective also 50% of the speed of light, but (3) from Erwin's perspective only 30% of the speed of light. This follows from the law of addition of velocities in SR, and this in turn from the "modified Pythagorean theorem" (Sect. 7.3). In contrast, the *spacetime distance* in SR is independent of the observer's perspective. It belongs to the *physical core of* the theory, which remains if one leaves the cloud of perspective dependencies behind.

Even more dramatic is the **Unruh effect of** QFT. Here it is shown that it depends on the reference frame what the vacuum is and what particles are. Where an observer in an inertial frame sees a pure vacuum, for an accelerated observer there is a flurry of particles whizzing around and vice versa. The whole confusing quantum world breaks upon us in this effect, and in a way that only appears in QFT, but not in QM. What we call "vacuum" or "presence of an electron" are in QFT already interpretations of certain states in an abstract state space. What such a state means to an observer depends on the measurements he can make of it, and these look different for an accelerated observer than for one at rest. Fortunately, this only becomes relevant at accelerations that are about 10^{20}-times higher than those to which we are typically exposed on Earth.

The third component of the cloud is the increasing **fuzziness** as we descend in the hierarchy of theories. In classical mechanics, we speak of particles having a unique location and a unique momentum. In QM, location and momentum (or at least one of the two) are subject to fuzziness, i.e., the state of the particle describes a *superposition* of different locations or momentums. But at least in QM it is still clear that there is a particle. In QFT, the number of particles present is itself subject to superpositions. For example, a state can be a superposition of a one-particle state and a two-particle state, such as when a neutron has decayed into a proton and an electron with some probability. The particles are excited states of a field, so are no longer independent "entities". Together with the Unruh effect, this completely breaks apart our notion of what "things" are, even the remnant of it that was left in QM. This is the subject of Sect. 9.2.

The Work of the Physicist

All these complications show that "facts" in physics are a decidedly difficult matter. Physical statements are always *approximations,* with a tremendous amount of *context* attached to them. To grasp this context is the whole art of the physicist, painstakingly learned in years of study and subsequent years of research.

A physicist who is supposed to say something about a given physical situation or solve a certain physical problem has to think about all kinds of things. First of all, he has to think about the theoretical framework in which he places the situation or the problem. Which scales play a role in it? In what order of magnitude are the effects to be considered? Which of them can be neglected within the framework of an *approximation* to so and so many decimal places? What simplifying assumptions can be made? When the physicist has decided, on the basis of these considerations, which theory is appropriate to the situation or problem, he usually has to carry out a calculation with the aid of that theory. In doing so, he must be constantly aware of which elements of this calculation arise from his *conventions* and what the "hard physical core" is that is actually at stake. He may in turn have to neglect terms in the course of the calculation in order to achieve further simplification, and may have to make additional assumptions in the process.

In the end, he succeeds in making a quantitative statement. Of this statement, which is formulated quite abstractly within the framework of the terms of the theory he has chosen, he must know how it *relates* to the physical situation, e.g. how an experiment or a measurement makes the quantity contained in his statement accessible to our senses, "translates" it into our sphere of experience. Finally, he has to evaluate the result with respect to its task. Of course, it depends on what this task actually was. Was it a matter of testing a theory that was not yet confirmed, or a confirmed theory in a range of scales where it could not yet be considered confirmed? Was it about making a quantitative prediction for the outcome of an experiment? Was it about "explaining" a phenomenon within the framework of the theory?

If the statement has not served its purpose, e.g. a prediction deviates from the actual outcome of an experiment, the task is to find out what needs to be modified. Was one of the simplifying assumptions wrong? Is there possibly a measurement error? Is one of the approximations made not accurate enough? Does the theory used not do justice to the problem? Is it even wrong? Can the theory be modified or extended without conflicting with 1000 other experiments? Or was only the interpretation wrong, how the theoretical statement is brought out in the experiment? So did the experiment end up measuring something *qualitatively* different than what was thought? For all these considerations, the physicist has to rely on his experience, the penetration of the whole context, as well as on his common sense.

In no way, however, do I want to give the impression that physical theories are nothing but "social constructs", as has been claimed in the course of the school of thought of **postmodernism** with regard to many sciences. If this were so, then almost every new experiment would have to give rise to a new theory. Each theory would only be part of the "social game" that physicists play with each other by inventing terms into which the experiments undertaken so far can be placed. Any new experiment would go beyond the scope of these "played" theories, unless they were so arbitrary that virtually any measurement could fit into them. In fact, however, all of physics is built on a few confirmed theories that are so successful that they make very accurate and correct predictions for almost any new physical situation that has not yet been played out; unless the situation is so complex that for *practical* reasons we are unable to perform the calculation that would give us the prediction. There are numerous checks and balances using a wide variety of observational methods to constantly ensure that the theories are correct and fit together. We can rely on these theories. All our technological achievements are based on them – from the light bulb to the mobile phone network, from the steam engine to space travel.

So even if the context attached to any physical statement is complicated, and, especially at the bottom of the hierarchy of theories, requires much experience, these theories nevertheless express **objective approximate truths.** Our intuition tells us that physics is about a reality that exists independently of human experimentation, and that the approximate truths of the theories relate to it in some way (in Duhem's sense; Chap. 6).

9.2   Things

In everyday life we deal with things that we perceive sensually. When we close our eyes and open them again, the things are still there. They exist independently of us, have a temporal continuity, even if nobody is looking at the moment. They have a certain shape, a certain weight, and a number of other properties that are also independent of whether anyone is looking at the moment. But above all, things *consist* of something, a **substance** or a mixture of substances, present in a certain quantity. "Substance" today is usually understood chemically; when we ask what something is made of, we usually get a chemical answer. Originally, the term comes from Greek philosophy, where it is characterized by some rather complicated definitions, but which ultimately go back to the fact that when we speak of properties, there must be a *something* that *possesses* those properties – a something that is present in a certain quantity, in a certain form, in a certain state. This something is what we imagine matter to be. The ideas we form about it are shaped by our sensory experiences. Matter can be seen, touched, sometimes heard, smelled and tasted.

If we go down the hierarchy of theories, there is not much left of it. The lower theories are about things that we can no longer see, touch, hear, smell and taste, abstract things in other words, but which, from the point of view of physics, are the basis for the things of our sensory perception. In QM, the properties of an object depend on the way in which we look at it. Things there seem to live in an abstract state space, and the question "state of what, actually?" has no clear answer. In other words, things there no longer have much to do with what we think of as things. Thus Arthur Eddington (1928) once wrote: *"We have chased the solid substance from the continuous liquid to the atom, from the atom to the electron, and there we have lost it."*

Everything we know is made up of these little elementary particles, and yet these elementary particles are quite different from anything we know. It's paradoxical. With QFT and the Unruh effect, even the particles no longer have an observer-independent existence. They are just mathematical expressions that result when you perform certain mathematical operations on an abstract state space.

We read a Feynman diagram of a scattering process of electrons like this: "Two electrons repel each other by exchanging a virtual photon." But this only means that we have distributed **nicknames** for certain mathematical expressions. What we have with the theory is an abstract space whose elements we nickname "state." States evolve according to certain mathematical rules as a function of a parameter, which we nickname "time". Some states perform only phase rotations (Sect. 7.6). The speed of rotation is nicknamed "energy". Among the states there is one with minimum energy, which we nickname "vacuum". We apply two certain mathematical operations to the vacuum and nickname the resulting state "two electrons". We study the time evolution of this state and nickname it the "scattering process". We realize that the best way to capture the scattering process is to use a certain computational procedure, via Feynman diagrams. Each diagram symbolizes a particular mathematical expression that represents a part of the time evolution. The mathematical expressions, in turn, are composed of smaller units, which we in turn give nicknames. One of them we call "virtual photon", and to form a whole sentence out of the nicknames, we say that this one has been "exchanged by the electrons".

A "thing" at this level is always just a nickname for part of a mathematical expression. The calculations that we perform with these expressions result in quantitative statements that we can relate to experiments. The mathematical expression nicknamed "electron" becomes the perceptible click or flash of a detector. The quantitative statements about the scattering process manifest themselves in the statistical distribution of detector clicks or detector flashes.

Ultimately, however, according to the theory, the detectors and our sense organs also consist of the same elementary particles. In theory, then, the fact that we see the detector flashing is only **nicknaming** for a mathematical expression involving the "exchange of photons" between a constellation of matter called "detector" and one called "eye". Schrödinger's dilemma (Chap. 2) takes on unexpected forms here.

Since reduction by replacement reduces everything sensual to the abstract, and the abstract is given in the form of mathematical structures, one can get the idea that mathematics is not only the language in which reality is written, but even its substance. No one has summed up this idea as clearly as Max Tegmark (2014) in his book *Our Mathematical Universe*. Physical reality, the universe, *is* a mathematical structure, he says. What we call "things" are substructures of this mathematical reality. We ourselves are also – in this view – partial structures of this mathematical reality.

If we ask ourselves "What is an electron?", the answer can only be a mathematical one. For the electron does not occur in our sensory world of experience, it is from the outset something abstract, the subject of physical theories that handle mathematical structures. These abstractions *refer* to measurements and experiments that we can perform, but the electron remains an abstract quantity in these references. Just because a detector clicks, it does not become sensually real, because it is not the electron that clicks, it only has to serve as part of an abstract explanation of the clicking. The same is true for all other elementary particles or quantum fields. Even apart from their mathematical abstractness, they have little in common in their behaviour, their properties, with what we understand as "things". So if we understand the universe on the level of these theories, we must conclude that this universe is a *universe without things*.

But maybe there is an "electron in itself", a "real", non-mathematical electron, which is *described* by the mathematical electron only in its quantitative properties? Sure, one can philosophize about such a thing, but it is difficult to justify. When Kant wrote of the *thing-in-itself*, he meant something that stands in contrast to our sensory perception or intuition. Kant pointed out that we perceive things in a certain way in space and time, but that we have no idea about how things actually are, because we can only ever refer to the forms of our intuition. Now the way in which we summarize the content of our perceptions in the form of "things" is already coupled to our intuitions. Should indeed to every cloud in the sky somewhere in reality beyond my perceptions belong the respective cloud in *itself*, to every wave of the ocean the respective wave *in itself*? It seems to me that the designation *thing-in-itself* is somewhat misleading, for it suggests a one-to-one relationship between the things of our intuition and the "real" things. Yet even within our intuition it is sometimes a matter of taste where one thing ends and another begins. Clouds have diffuse boundaries, and sometimes you will see two clouds where someone else sees only one. I rather understand Kant to mean a *reality in itself* about which we cannot say anything, not even into what kind of "things" it can possibly be broken down.

So we have several levels to distinguish:

1. The underlying reality, of which we cannot say what things are in it (and whether "thing" or "entity," as the philosopher puts it, is even an applicable concept here)
2. The world of our perception, which is shaped by our ideas fed by sensual experience; that is the world of things.
3. The mathematical abstractions that we *apply in* the context of physics at level (2)

The electron is at home on level 3 and there it is part of a mathematical structure that helps us to effectively describe observed phenomena on level 2. Now, moreover, the structures at level 3 are only to be understood as approximations which together form a hierarchy of theories. The electron in QM is (mathematically) something different from the electron of QFT (in the former case an individual quantum object, in the latter case an excited state of an electron quantum field). In terms of Duhem's *natural classification,* it is plausible to believe that both variants of the electron *reflect* certain aspects of reality. But to assume therefore that in reality, i.e. at level 1, there is a "thing" worthy of the name electron because it is in a one-to-one relationship with the mathematical expression nicknamed "electron", but which is not itself anything purely mathematical but is merely *described* by the mathematical electron, I think is a vague speculation that is not justified by anything.

9.3 World and Reality

Let us stay with the three levels just mentioned, (1) reality, (2) world, (3) physical theory, and try to sort out their relationships to each other. Reality is that which underlies everything and from which we and all phenomena ultimately emerge. By the world, on the other hand, is meant the world of experience, that which we make accessible to ourselves through our sense impressions and our thinking. It is thus first of all an *idea;* an extrapolation of what we perceive and think we know; an idea that always runs implicitly in the background of our consciousness.

The world appears to us as a *common* world, about the properties of which we can *agree* to a large extent by means of the method of natural science. Natural science, as shown in Chap. 5, is reductionist in structure and thus acquires a hierarchical structure, at the base of which is physics. Physical theories, then, form the foundation of what we *know* about the world in terms of natural science. They form a *picture of* the world written in the language of *mathematics*. This image implies a complicated system of translations (Chap. 6) that establishes a relationship between the sensually perceived events in an observed phenomenon or a physical experiment and the mathematical expressions from which the theories are formed. This works so well that, in Duhem's sense, we sense the theories of level 3 to be a reflection of actual relations of reality (level 1) (though it should be emphasized that a reflection is something quite different from an equation).

This essentially sums up the relationships of the three levels. If one remains modestly within this view, even QM does not seem so dramatic. In fact, from my point of view, it seems to insist on maintaining the separation of the three levels, as already shown in Sect. 7.6.

However, the most common view among both philosophers and natural scientists at present is **physicalism**. Roughly speaking, this can be expressed as "everything is ultimately physics." Put more precisely, physicalism is a metaphysical *position* that can be summarized in two statements: First, it says that reality is to be understood in terms of *an ontology*, that is, in terms of a compilation of "everything that exists," a list of existing things (or *entities*, in philosopher's jargon). Second, it says that these existing things are precisely the things that physical theories are about.

However, this approach is inconsistent with the relations presented above: things are at home on level 2. The decomposition of the world into things moving in space and time is part of the way our idea called world works. Neither is it clear that the underlying reality can be decomposed into "existing things", into entities, nor are the fundamental physical theories about any things. After all, we have just seen that everything that looks like a "thing" there is really just a nickname for a mathematical expression. So at level 1 the concept of *entities* is extremely questionable, at level 3 it is just plain wrong.

Physicalism in its conventional form therefore makes no sense at all in my point of view. It only expresses a great misunderstanding. However, it can be saved in only one way: namely, by recognizing mathematical structures and their substructures as entities themselves and granting them an independent existence in the sense of Platonic philosophy. Plato described in his famous allegory of the cave that everything earthly is only a shadow image, a reflection of true and eternal ideas, which have a timeless-eternal existence in the "heaven of ideas". Thus, if one wants to save physicalism, one must postulate that mathematical structures exist in a kind of *Platonic heaven of ideas*. Second, one must postulate that this heaven of ideas is reality, and indeed *all of* reality.

Reality is the totality of mathematical structures. Third, one must postulate that there is an all-encompassing physical theory that describes our entire world of experience (level 2), that is, all observations and experiments ever made, in terms of a single mathematical structure (level 3). Then one would have to conclude (because one has already postulated that nothing else exists except mathematical structures) that our world of experience *is* in fact nothing else than this structure. That is, we ourselves are mathematical substructures of it, and all our observing and experiencing and experimenting come only from how each of these substructures is embedded in the overall structure, how it participates in it, in particular in which – purely mathematical – relations it stands to other substructures.

This form of physicalism seems extremely daring to me, but it is at least consistent, free of conceptual nonsense. It is exactly the position of Max Tegmark (2014), as he describes it in his book *Our Mathematical Universe*. As daring as this view is, it has a certain appeal. Traditionally, the empirical path of natural science has always been highly critical of idealistic thought such as that of Plato. The mathematical universe, however, brings the two closely together.

In his *history of philosophy* Hirschberger writes[1]: *"There are really only two pure types in philosophy: Plato and his antipode David Hume. Everything else can be apportioned to one of each or is a mixed type."* Whether it's really that simple with philosophy, I don't want to judge here. But let us follow Hirschberger's thesis for a moment. Plato stands for the pure ideas, the "logos" that unfolds in them. The material world is only a shallow reflection of these ideas. This can be seen from the fact that the world is **contingent**, that is, arbitrary; things could be one way or another. In the world of ideas, however, everything is as it *must be,* it represents the actual reality. Hume stands for pure empiricism, the basis of the natural sciences, which seeks truth precisely in the "contingent" experiences of the material world, concludes from them as much as can be concluded, and dismisses everything else as speculation. It is to experience, not to ideas, that our truth is bound, according to Hume. Tegmark's view brings together the two so different philosophies: Empiricism, the method of the natural sciences, leads us to the laws of nature, which can be reduced further and further until we finally arrive at a theory of everything. This theory is contingent only insofar as we can ask: Why is there a world with *these* laws of nature, surely there could have been others? This last vestige of contingency is removed by the fact that the world described by the all-encompassing theory is nothing but a mathematical structure that *lives together*

---

[1] [Hirschberger, 1980, vol. 2, p. 652].

with all other conceivable mathematical structures in the Platonic heaven of ideas. The "pure type of Hume" is thus ultimately traced back to the "pure type of Plato." What a grandiose unification of philosophy!

So postulating a purely mathematical reality seems to me to be the only consistent variant of physicalism. On top of that, this variant has a special philosophical appeal. On the other hand, we have to note that it is an extremely daring, purely *metaphysical* position, which cannot be proven at all, which can at most be found plausible. But do we find it plausible? That is for each person to decide for himself, as it is in general the case with metaphysics.

Metaphysical positions cannot be proven, but some can be refuted if they are logically inconsistent or contradict the findings of natural science. The hypothesis of mathematical reality, however, is not one of these. However, I still think it is false and we can also *see that* it is false. It does not contradict anything as objective as logic or natural science. But it does contradict what we can *subjectively* see. This statement requires some further explanation and extensive discussion. It is the subject of Chap. 11. If you are not convinced by it and "subjective cognition" remains suspicious to you, it is open to you to continue to believe in a purely mathematical reality without running into a logical contradiction.

However, if we reject Tegmark's hypothesis, but continue to believe with Duhem that we reflect *aspects of* reality with the theories of physics, we arrive at a multi-layered worldview. In it, each of the proven theories takes on a certain role, represents to us a certain partial character of reality, but without these theories, even in their totality, fully grasping reality, firstly because they all represent only approximations, and secondly because reality also has other, non-mathematical aspects which cannot be found in the theories.

9.4 Time

For physics, time is still mysterious. Perhaps only as a physicist can one appreciate *how* mysterious time actually is. We will provide some order here by distinguishing four different aspects of time and summarizing what physics has to say about each of them. The four aspects are as follows:

1. **Linearity:** The linearity of time allows us to sort events into a certain **order** by arranging them along a straight line, the *time axis*. What is meant here is initially an *undirected* order, i.e., we do not yet want to commit ourselves to which direction to read the time axis. If we arrange four events, A, B, C and D, on the time axis in this order, we can understand the resulting

sequence as ABCD or as DCBA, depending on which way we look at the time axis. What is clear in both cases, however, is that (1) B lies between A and C, and (2) C lies between B and D. In all theories except statistical mechanics (and in the context of certain interpretations of QM), the laws of physics are time-reversal invariant, i.e., any process is reversible, i.e., can proceed forward just as it can proceed backward, hence this restriction. To describe such reversible processes, the undirected order, linearity, is already sufficient. Linearity is unproblematic in Newtonian mechanics, where time has an "absolute" character.

In SR, time is "relative", but the significant fact that nothing can move faster than the speed of light, that is, that everything moves along *time-like* lines through spacetime, causes the sequence of events to be unambiguous, provided those events are in some contact with each other. Time remains linear. If we could move through spacetime at faster-than-light speeds, that is, along spacelike lines, the order of events would depend on the perspective of the observer. Erwin could throw a faster-than-light ball to Otto, and Otto could throw it back in such a way that it would arrive at Erwin before he had even thrown it off to Otto. Erwin would then have the same ball in his hand twice for a moment. The whole concept of time would be completely unhinged – abysses of paradoxes would open up. So let's be glad that faster-than-light speed is not possible.

In GR, there are indeed spacetime structures with closed timelike lines, i.e. solutions of Einstein's field equations that admit possible observers for whom time runs cyclically, like Groundhog Day for Phil Connors. This too would allow temporal sequences to be rearranged, with paradoxical consequences. So let us be glad that our universe does not seem to conform to any such solution.

2. **Duration:** In order to be able to define the duration of a period of time, we need a *clock,* a comparative measure (similar to the definition of the meter; Sect. 9.1). Such clocks are abundant in nature through uniformly periodic processes. Thus the rotation of the earth around the sun defines the year, the rotation of the moon around the earth was the origin of the month, and the rotation of the earth around itself determines what a day is. This allows us to make calendars, celebrate birthdays, and tell our age in years. By further dividing the day into hours, minutes and seconds, we get clock times. To measure these, we are helped by clocks based on other uniformly periodic processes, such as the swinging of a pendulum. The laws of physics favor the occurrence of such processes; the universe has rhythm in its blood.

By comparing the date and time of different events, we can determine the time span, i.e. the duration, that lies between the events. Once again, SR and GR complicate the matter by showing that the time span between two events depends on the reference frame. But related to a given frame of reference, duration remains a well-defined, unambiguous concept.

3. **Direction:** The direction of time implies that we can distinguish earlier from later, or past from future. We remember the past but can only guess at the future. The direction of time is also a basic requirement for the concept of causality. For only if it is clearly defined that an event A occurred earlier than an event B, are we justified in the hypothesis that A is the cause of B and B the effect of A. If, on the other hand, the sequence were arbitrary, cause and effect would also be interchangeable, and causality would be a meaningless concept.

   Although the distinction between past and future seems to be completely self-evident for us in everyday life, an unambiguous time direction is interestingly nowhere built into the fundamental theories of physics. It is only made possible by special *solutions of* the theories that exhibit a certain temporal asymmetry in the sense of statistical mechanics, a state of low *entropy*, as discussed in Sect. 7.5. The considerations involved are very complicated and involve several other fields besides statistical mechanics, such as cosmology and QM. It is certainly one of the most fascinating and complex problems in physics. Physics has a great deal to *say on* this subject, to contribute thoughts that philosophy would never have come up with by pure thought.

   An excellent popular science account of these considerations can be found in Sean Carroll's (2010) *From Eternity to Here.* For experts with a degree in physics, *The Physical Basis of the Direction of Time* by Heinz-Dieter Zeh (1989) is an excellent general overview.

4. **Passage:** Time passes. We move "through it, away from the past, towards the future". The vehicle on which we move is called the Now. We are always where the Now is, and the Now leaves only scorched earth: every moment it once passed is lost forever, is no more, will never be again, only in our memory. This character of time is essential to us, it makes up the drama of our lives. But it can apparently only be expressed poetically. As soon as one tries to even *define* the passing of time within the framework of a mathematical language, so that one could at least check how it might fit into existing or future physical theories, one encounters insurmountable obstacles. In the theories of physics, time simply does not *pass*; it is not even possible to express what this passing might even mean. Physics has nothing at all to say about this.

This has now led some to declare the passage of time to be an illusion, a psychological effect similar to an optical illusion. Others, such as Lee Smolin, think that it might be possible to find a way to introduce the passage of time into physics, and are searching feverishly for it. Still others, including Einstein, think that the passage of time is an important aspect of reality that physics simply cannot talk about. I would like to join Einstein's faction here. The topic deserves a somewhat more detailed discussion, which will be done in Chap. 11.

Thus, it turns out that the aspects of linearity and duration of time fit quite naturally into the existing theories of physics, except for some complications brought about by SR and GR. The direction of time, on the other hand, is an extremely complicated problem that poses great challenges to physics, but is at least treatable within its framework. The passage of time, on the other hand, seems to be entirely outside of physics.

10

The Practical Limits of Physics

We have already talked about the *practical limits of* physics, but here we want to give another overview, together with a closely related topic, the *scale dependence of* physics.

By practical limits I mean: there are areas of physics or the universe that are not accessible to us. That is, there is something here that physics has something to say about *in principle,* but because of the limited possibilities available to us as humans, we can't figure it out.

These limited possibilities have several causes: Some are related to natural laws that exclude us from observing certain regions of the universe - spatial or temporal - from the outset. Others are related to the limited resources at our disposal, which prevents us, for example, from building arbitrarily large particle accelerators.

The first category includes the regions of the universe that are hidden behind a *horizon.* The largest of these horizons is the *cosmological horizon,* the boundary beyond which light has not had enough time to reach us since the beginning of the universe. This boundary is indeed shifting further and further away from us with time, but at the same time the universe is also expanding, it is even expanding faster and faster, and this effect is to be reckoned with the continuous receding of the horizon. This makes it look quite likely that much of the universe will remain hidden from us *forever.* Except that humanity probably won't exist for all eternity (nor will the intelligent entities that may follow it and carry on its science), so it doesn't do us any good to know that a region will come within reach of our telescopes in a trillion years. The question of whether the universe is finite or infinite is thus most likely beyond our grasp.

© Springer-Verlag GmbH Germany, part of Springer Nature 2022
J.-M. Schwindt, *Universe Without Things,*
https://doi.org/10.1007/978-3-662-65426-2_10

Another type of horizon are the *event horizons* that surround black holes. These ensure that the only way for us to explore the inside of a black hole is to plunge in, never to be seen again. Much of what it looks like inside is known to us from theory, GR and QFT - theories that are well backed up from other observations, but firstly it would be nice to have these predictions confirmed by observations in this exciting case, and secondly there are also unanswered questions, especially those related to the singularity at the centre of the black hole.

There are other constraints in the field of cosmology. We would like to understand the temporal evolution of the universe from beginning to end. The further we look, the further we look into the past, because of the duration of the transit of light from the source to our telescope. So, first of all, the problem arises that we can only cover any period in the history of the universe with data from sources of a very specific distance from where the light took just the "right amount of time" to get to us. This problem is not too great because, by all evidence, the universe is quite homogeneous in terms of its large-scale structures. It seems to have evolved equally everywhere, so we can infer the past of all other regions, including our own, from data coming from one distance layer.

A much bigger problem is that the universe only became transparent about 400,000 years after the Big Bang. Now, many of the open questions in cosmology relate precisely to the phase shortly after the Big Bang: How did the excess of matter over antimatter occur? Did an inflationary phase of the expansion take place? Or how else did the strong homogeneity over large distances come about? How and in what context did the Big Bang itself occur? Was there a before or a beyond of it, as some models of inflation suggest? Speculative theories exist on all of this, but we can't test them because this phase is literally in the dark. Some circumstantial evidence can be read from the cosmic background radiation, but it's just circumstantial evidence, traces from a time long after the events we're actually interested in. Perhaps we can gather some more clues with the help of neutrinos or gravitational waves, but it seems doubtful whether this will clear up the big questions.

We can look into the past with telescopes, but not into the future. From the confirmed theories, we can predict the "near" future of the universe's evolution with a high degree of probability, i.e., the next billions, possibly quadrillions, sextillions, and even more years. (Hard to say at what point it seems "likely" that something unexpected will happen; the events involving galaxies and even larger structures are so much more predictable than our fickle life on Earth). But the fate of matter as a whole in the even more distant future depends on laws of physics that we haven't yet sorted out, especially the question of whether protons will eventually decay, albeit with huge half-lives, into other particles after all, as many unified theories predict. Another imponderable is the so-called dark energy. Does it behave for all time as it does now, i.e. as a constant vacuum energy density, and thus lead to an eternally accelerated expanding universe, or will this behaviour change at some point, possibly even to the effect that the universe contracts again? The distant future can be fathomed with theories, but these contain remaining uncertainties which we cannot clarify by observations, because we cannot look into the future.

The further away the objects we can see with telescopes, the less sharp we can resolve them. We can't make out details. Impossible to determine if there is a blue planet like Earth in a galaxy 100 million light years away, and what its inhabitants look like. The details far away remain hidden from us. We can't do experiments there either, but rely on the dull light (and other forms of radiation) that reaches us from there and from which we can draw our conclusions (an astonishing number, it must be said).

Our ability to travel there is severely limited, physically and technically. Perhaps at some point we will succeed in interstellar space travel to our closer galactic vicinity (at the moment we first have to make it back to the Moon and then to Mars), to star systems several light years away. Maybe we'll manage to send unmanned probes even further. Some science fiction writers dream that one species could even colonize an entire galaxy. But if we remain realistic, travel to celestial bodies several hundred light years away, and even more so to galaxies millions of light years away, is as good as impossible.

The mysterious dark matter contributes a large part of the mass of our Milky Way and also of all other galaxies. We can observe its gravitational effects, mainly through the rotational speeds of galaxies and its effect on the expansion of the universe. But these effects are too unspecific to say what properties this form of matter has apart from its total mass, whether or not it is made up of particles that can be classified within the framework of a QFT and, if so, what kind of fields these particles arise from. We have not been able to produce particles in our experiments with particle accelerators that could be candidates for dark matter. This could simply be because this form of matter only interacts via gravity with the "normal" matter we can work with in such experiments, but not via the other interactions the latter is subject to, i.e. the electromagnetic interaction and the two nuclear forces. If so, it will be hard, perhaps impossible, to find out anything more precise about them, since they escape our detectors and do not show up in the cascades of particles we produce in our elementary particle experiments. It would then have nothing to do with the lack of resourcefulness of physicists that the details of dark matter remain uncertain, but simply with its inaccessibility resulting from a lack of interaction. We might then continue to know of it only through its effect on galaxies and the large-scale structures of the universe. Perhaps it even consists of several different types of particles, which possibly interact violently with each other, but of which we do not notice the slightest, because we are not involved in these interactions. Or it does not consist of particles at all, but emerges from something completely different, completely unknown to us.

Equally inaccessible are the high energy scales of Grand Unified Theories (GUTs) and quantum gravity. Even the LHC particle accelerator, which costs many billions of euros, can only bring particles to energies of a few teraelectronvolts (TeV; trillions of electronvolts), which is about a trillion (10^{12}) times too few to test the expected effects of a GUT, and a quadrillion (10^{15}) times too few for quantum gravity. This large gap is unlikely to be bridged.

So far, these were all practical limits arising from our limited ability to observe and experiment. But the theory side of physics also has practical limits, namely those arising from complexity. Our limited computational capacity - even with the largest computer clusters, it remains finite after all - means we can't do many calculations we'd like to do. Sometimes there are too many variables involved (sometimes even infinitely many), sometimes the differential equations are too difficult, often we are dealing with "nonlinear" phenomena where tiny inaccuracies inflate exponentially over time, so that we would have to compute with infinite precision to make any reliable prediction at all. The extent of these difficulties depends on the specific problem.

Often, however, it is a matter of applying known theories to complicated conditions. For most theories, simple special cases can be constructed in which the calculations are comparatively easy. By means of such special cases the statements of theories can be sufficiently concretized and tested. Therefore, the problem of complexity does not, in general, stand in the way of physical theory construction, but of its later application. Thus, when we think of the practical limits that prevent us from progressing in fundamental physics, at present it is rather the above-mentioned limits of observation and experimentation that hold us back, not so much the computational problems in theory.

10.1   Once Again: Scales

A large class of our practical limits can be understood in general terms in the context of the scale-dependence of physics. Let us therefore summarize once again: Different laws of physics are relevant depending on whether we are studying phenomena at the atomic level (scale of picometers, i.e. trillionths of a meter), at the human level (scale of meters), or at the cosmological level (millions to billions of light years). Similarly, it makes a big difference whether we analyze phenomena that take place in tiny fractions of a second, on human time scales of days and years, or whether we study the history of the universe, which takes place in billions of years. Also relevant is whether or not the velocities involved in a phenomenon are close to the speed of light. In particle experiments, energy and momentum also play a major role. Particles with very high energy behave very differently from those with very low energy.

The different kinds of scales (space, time, energy, velocity or momentum) are not independent of each other. SR establishes a connection between space and time, seconds can be converted into meters (if one uses the "natural" convention $c = 1$; Sect. 7.3). In quantum theories (QM and QFT) there is also a connection between momentum and length: the higher the momentum of a quantum object, the smaller its wavelength. This wavelength is also, very roughly speaking, a measure of the distances that a microscope using such quantum objects can just about resolve. Finally, it is also true for highly relativistic objects, i.e. objects moving almost at the speed of light (as is almost always the case with particles shot at each other in particle accelerators), that their energy is almost identical to their momentum. Thus, via the confirmed physical theories, space, time, momentum, and energy scales are directly related; they can be converted into each other, with a small space or time scale belonging to a large energy or momentum scale. The **Planck scale**, for example, the scale on which the effects of quantum gravity are expected to make themselves felt, can be given as an energy or momentum scale with 1.2×10^{19} GeV, as a length scale with 1.6×10^{-35} m, or as a time scale with 5.4×10^{-44} s.

Thus, at the fundamental level, we can reduce all these scales to length scales, which are the most descriptive for us. (The particle physicist, on the other hand, prefers energy or momentum scales, since his calculations are often performed in terms of energies and momentums). So if we want to situate all known physics in terms of length scales, it ranges from 10^{-20} m, the scale reached by the high energies of the particles produced at the LHC, to the distance of the cosmic horizon, about 10^{26} m. It therefore covers about 46 orders of magnitude. Since we know how much the relevant physical laws in each case change as one zooms back and forth between these scales, there is no reason for us to expect the changes to stop outside the section we can access. For example, we expect substantial changes at the Planck scale, 10^{-35} m. At the other end, most cosmic inflation models reckon that the visible part of the universe is just a tiny grain in a much larger region of space, a kind of bubble that has inflated within even much larger spatial structures.

The accessible range from 10^{-20} m to 10^{26} m is a finite section on a basically infinite *line of scales* (Fig. 10.1). There is no guarantee that the changes will

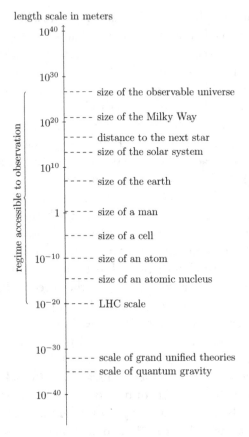

Fig. 10.1 Finite section of an in principle infinite "scale line". The LHC scale refers to the length scale that is just resolved by the LHC particle accelerator

stop anywhere on this line. So it may be that the practical limits imposed on us mean that we know only a finite number of physical laws (or theories) in an infinite hierarchical structure, so that at any given time we know very much more not than we do.

However, this is not necessarily so. There is a possibility that the line of scales is limited at both ends. The universe could be finite, and then its size (or diameter) would be the largest length imaginable. One can also speculate that space or spacetime is granular, "pixelated", that is, that there is a smallest distance that cannot in principle be resolved further. Or it could be that the line of scales is infinite, but beyond certain limits no new physical laws come into play. This would be the case if there is a theory of everything from which all other theories can be derived. Because in general the laws at small length scales determine the laws at the larger length scales, not vice versa, one reckons that such a theory is defined at small length scales, below the accessible range, and all known theories at the scales accessible to us are then *effective theories* arising from the fundamental one. Finally, there is also the possibility that the concept of length itself is a derived, emergent concept, and that at some point one arrives at a level where the concept of scale itself loses its meaning.

The point is, we don't know all this and can only speculate. None of the practical limits are absolutely set in stone, it's too early to give up, on such an endeavor. Maybe we'll get lucky and find that the universe is finite. Or maybe there is an interaction between dark matter and "normal" matter that we can work with after all. Maybe we'll come up with a theory of quantum gravity that we can somehow confirm on the scales we have access to, and that also guarantees us a minimum length, a lower end of the line of scales. All this has to be tried further.

But we also have to be realistic and face the very plausible possibility that we will be stuck with many of our unanswered questions because, due to the practical limits of physics, the answers are beyond our grasp.

11

The Fundamental Limits of Physics

In the previous chapter, we examined the practical problems that limit our search for physical laws and prevent us from completing the physical world-view at all ends. In this chapter, we now ask whether there are phenomena or aspects of reality that cannot in **principle** be grasped by scientific methods or, more precisely, that do not fit into the scientific picture because they are "orthogonal" to it in some sense yet to be specified.

Some will think first of questions of ethics, aesthetics, or religion (the good, the beautiful, and the sacred). But an ardent supporter of the omnipotence of the natural sciences might argue that these three things can be understood as purely sociological or psychological constructs, purely human conventions that serve a particular purpose.

We have already discussed another point, which certainly lies outside the realm of natural science: The question of what is real escapes the methodology of natural science. Linked to this is the question of why anything exists at all. However, these questions are very abstract. Can we find something more concrete, something that we have before our eyes every day, of which we can say with certainty that physics has nothing to say about it?

This chapter will not be about ethics, aesthetics or religion, but about our **experiencing**, about the way we subjectively experience the world (or whatever). To do this, we need to have a look at the **philosophy of mind**, its concepts and its various positions on the subject. There, the question of to what extent and in what sense our experiencing can be traced back to brain processes has long been debated. This question is considered by many to be decidedly difficult, and is therefore also known as **"the hard problem of consciousness."**

© Springer-Verlag GmbH Germany, part of Springer Nature 2022
J.-M. Schwindt, *Universe Without Things*,
https://doi.org/10.1007/978-3-662-65426-2_11

I will side with those in the discussion who say that experiencing does not fit into the scientific picture at all, that brain processes cannot produce experience, and that no extensions of physics or biology or our understanding of them can change anything about it. Thus we have a phenomenon or aspect of reality that we can directly recognize (it is not speculation or a matter of faith) and that lies outside the fundamental limits of science.

An important aspect of our experience is that time passes in it. Our experience rides like a surfer on a wave called the now, moving continuously through time at a rate of one second per second. This wave, this Now, and therefore the passing of time, does not occur in physics. There is time in physics, and even an "arrow of time", but time does *not pass* because there is no Now there to move along the arrow of time. Thus, the passing of time is another phenomenon that lies outside the fundamental limits of science.

These things are controversial though, I don't expect to get agreement from all sides here.

11.1   The Hard Problem of Consciousness

The concept of consciousness is quite central to the philosophy of mind, but has also led to some confusion because it is used in very different ways. Just read through the enumerated meanings on Wikipedia, it can make one dizzy. To avoid such confusion, Ned Block (1995) introduced the distinction between access consciousness and phenomenal consciousness. Access consciousness is about being aware of a fact or process in the sense that one can access it mentally or linguistically, i.e., express oneself about it, report it, and incorporate it into one's thought processes (as opposed to unconscious processes where we cannot). Phenomenal consciousness, on the other hand, is our experience, i.e., our subjective inner perspective of "how something feels" or "what it is like to be me in the moment" or how something presents itself to our "inner eye" or "inner ear."

The "hard problem of consciousness" refers exclusively to phenomenal consciousness. Access consciousness is somewhat easier to address because it is much easier to experiment with. The ability to express a fact linguistically or by gesture is objectively testable, but an inner experience is not so easy, unless the inner experience is expressed linguistically or otherwise outwardly, in which case access consciousness is again involved. Moreover, access consciousness clearly goes hand in hand with certain cognitive abilities, which brain research is getting better and better at locating and tracing in the brain. With phenomenal consciousness, on the other hand, the connection with cognitive processes is rather unclear.

The discrepancy between the two types of consciousness can be illustrated particularly impressively in so-called split-brain patients, in whom the connection between the two hemispheres of the brain is destroyed. The left hemisphere of the brain – sensory and motor connected with the right hemisphere of the body – is responsible for linguistic expression and rational thinking. Such a patient cannot express what is going on with his left side of the body (or even on the left side of his visual field). His access consciousness has no access to it. But does this mean that he does not feel anything on the left side? Probably not. Does he then have *one* phenomenal consciousness or *two separate* ones? Hard to say.

In the sense of the "hard problem of consciousness", the question of how Beethoven's brain composed his symphonies is not an easy one; but the question of how it is possible that in the process he virtually listened to the music on his "inner ear" is much more difficult. For composing is "only" an activity of access consciousness. The tone sequences are put together in a conscious process (even if this may have been preceded by some unconscious preparatory work) and symbolically notated. The inner experience of these melodies, however, is a process in phenomenal consciousness. Meanwhile, computers can also be made to compose simple pieces. However, hardly anyone assumes that this is accompanied by an "inner experience" on the part of the computer.

In his influential article "What Is It Like to Be a Bat?" Thomas Nagel (1974) gets to the heart of the problem with subjective experience. He explains that even if one were to understand the perceptual system and brain processes of a bat in all details, one would still not know what it is like to be a bat. The phrase "what it's like to be …" as a paraphrase for subjective experience has since been used as a fixed term in numerous philosophy of mind texts *("the what-it's-like-to-be")*. The problem is that "what it's like to be X" can only be experienced from X's point of view. In natural science we try to objectify, i.e. precisely to eliminate anything specifically subjective about a state of affairs. Now how is natural science supposed to be able to describe and explain, of all things, the subjective? Nagel writes:

> [...] that every subjective phenomenon is essentially connected with a single point of view, and it seems inevitable that an objective, physical theory will abandon that point of view [...] If the subjective character of experience is fully comprehensible only from one point of view, then any shift to greater objectivity – that is, less attachment to a specific viewpoint – does not take us nearer to the real nature of the phenomenon: it takes us farther away from it. (Nagel 1974)

Nagel chose the bat as an example because it is developed enough that we can assume that it has something like a subjective experience, but on the other hand it is also sufficiently different from us in its perceptual apparatus (it perceives the external world primarily by means of ultrasound or echolocation) that we cannot so easily "put ourselves in its place". With another human being we could say: "I can imagine what it is like to be you; I can put myself in your place, because your perceptual apparatus is the same as mine, you see the same colours, hear the same sounds. I know your situation, I can imagine the feelings you're going through because I've experienced similar things myself." You can't do that with a bat.

Further, Nagel expresses his conviction that there is something about subjective experience that is in principle beyond the reach of human concepts. He accuses materialism of simply ignoring the problem:

> *It is useless to base the defense of materialism on any analysis of mental phenomena that fails to deal explicitly with their subjective character. [...] And to deny the reality or logical significance of what we can never describe or understand is the crudest form of cognitive dissonance.* (Nagel 1974)

The debate about the significance of Nagel's article has since been conducted with similar intensity and variety as the debate among physicists about the correct interpretation of QM.

For the question "Why do certain organisms have subjective experience?", David Chalmers (1995) has coined the term "the hard problem of consciousness". Chalmers contrasts this with some other problems that are not exactly easy either, but are easier than the "hard problem" because at least rudiments of a scientific approach can be identified, for example: How does the cognitive system bring together the information from our various sensory perceptions? How does it classify, categorize, and respond appropriately to these perceptions? How does the brain manage to focus our attention on something? How does the difference between sleep and wakefulness occur? These questions relate to specific abilities or to the performance of specific functions, whereas the "hard problem" is about the subjective feelings that go along with all of this.

Qualia (singular: quale, from Latin *qualis*) are a central concept in the discussion of the "hard problem". Qualia are the "how it is for us subjectively" of our perceptions.

Let's take the following situation for example: I see a red traffic light and step on the brake. The objectifiable part of this situation is like this: The red light hits my retina, the information about it is transmitted via the optic nerve to the brain, brought together by cognitive processes with the other details of the situation and processed. This leads to a reaction: the muscles of my right leg are instructed to make a small movement up, then left, then back down; I've switched from the accelerator to the brake. Somehow, though, this information processing goes hand in hand with the red glowing circle in my phenomenal consciousness that represents the red light of the traffic light, as part of a bigger picture. The red on the "inner eye" or "inner screen," or whatever you want to call it, is the quale that corresponds to the perception of the color red – a perception that, when viewed from the outside, presents itself as a neural pattern of excitation in the brain. Personally, I quite like the "screen" as a metaphor. We speak of an "inner screen", but this screen is of course not located as such in the brain. "Inside" here is meant in the sense of subjective, not accessible to others. The "hard problem" then is: why does this "inner screen" exist at all? It contributes nothing to information processing. The cognitive recognition of the red light and the stepping on the brakes (access consciousness or even partially unconscious recognition and action) occur regardless of whether it exists or not.

Qualia are used for various thought experiments on the basis of which the "hard problem" is discussed. Three experiments are particularly popular:

1. **Qualia inversion:** In qualia inversion, the qualia of two people, let's call them Alice and Bob, are reversed in comparison to each other. The quale that appears with the color red for Alice appears with the color green for Bob, and vice versa. Both respond cognitively identically to the red light, i.e., they recognize the color as red and slam on the brakes, but in phenomenal consciousness (on the "inner screen") the colors are reversed. We say "are reversed in comparison to each other", but in fact Alice and Bob cannot compare. The "screens" are private, both have never seen anything but their own, they can never find out about the inversion. Using this and also the following thought experiments, the philosophy of mind discusses the questions of whether qualia exist at all, whether situations like this can be imagined, and, if so, what the consequences are. The reactions are very different. For some it makes sense, for others it doesn't. I remember a discussion about perception in my school days in which a classmate asked a teacher, "Can't it be that your red is green to me and your green is red to me?" I guarantee the classmate had never read anything about qualia inversion, the thought just came to him spontaneously. I knew immediately what he meant, although I too had never read anything about qualia and qualia inversion. But the teacher didn't know what to make of it, "Well

then you'd be in a lot of trouble on the road." - "No, you don't understand what I mean."

2. **Zombies:** Philosophical zombies are people who have only access consciousness, but no phenomenal consciousness. They lack qualia, but behave like normal people. Let's say Bob is a zombie, but Alice is not. Again, both process the sight of the red light in exactly the same way and brake, except that Alice additionally experiences the whole thing on her "inner eye", but Bob does not; he acts "like a robot", without a subjective inner life, but still exactly as if he had one. As with inversion, Alice can never determine that Bob is a zombie. Is such a thing conceivable? Does that include the possibility that all humans but me are zombies? And for that matter, how am I so sure I'm not a zombie myself? Does it follow that the whole notion of qualia is nonsense? Or does it just follow from the conceivability of zombies that qualia exist and are not reducible to brain states?

3. **Mary:** The Mary thought experiment is really called the Mary thought experiment, as if it matters that Mary is called Mary. The example is a bit reminiscent of Nagel's bat. Mary has never seen colors in her life, she's been locked in a black and white room since birth. But she's read everything there is to know about colors (including that red lights mean you have to brake) and how the eye and brain accomplish color vision. Then one day she steps outside for the first time and sees colors. Question: Does she learn anything new? The author of the thought experiment (Jackson 1986) thinks, first, that the answer is clearly yes and, second, that this disproves materialism. Both, of course, were subsequently challenged by colleagues.

The whole qualia issue somehow seems to be a red rag that provokes strong reactions. In discussions I have almost always experienced either vehement rejection or clear agreement, rarely a middle ground, a cautiously critical but open attitude. Perhaps the rejecters are the philosophical zombies, after all, and that's how you can tell. It would be a great irony if it were the other way around.

11.2   The Flow of Time

A very special aspect of our experience is that there is a **now** and time is passing. Rudolf Carnap remembers his conversations with Albert Einstein:

Once Einstein said that the problem of the now seriously worried him. He explained that the experience of the now meant something special for man, something essentially different from the past and the future, but that this important difference did not occur in physics and could not occur there. That science could not grasp this experience seemed to him an object of painful but unavoidable resignation [...] There was something essential concerning the now that was simply outside the realm of science. (Carnap 1993)

What exactly is the difficulty Einstein is talking about here? We experience time from the perspective of a now that divides time in half and that seems more real to us than the past or the future. The past seems "lost forever" to us, the future is "not yet here". Worse, the now seems to move through time toward the future (time "passes"), or, equivalently, time moves through the now toward the past, so that the amount of what is "lost forever" keeps increasing.

In physics, however, neither a distinguished now nor a movement of it occurs. Physics describes the world on the basis of mathematical structures, and these are "simply there", as a whole, without any change happening.

In GR this is called a block universe. The whole space-time is one single structure, it is given "at once". The meaning of past and future in it is not much different than that of north and south. We can indeed imagine that in it something runs from "south" to "north". But this idea is not justified by anything in physics. Hat and van Fraassen write:

> *Our experience of time is that of a moving present, one that is very different from a tick mark somewhere halfway down a time axis in a frozen four-dimensional space-time.* (Hut and van Fraassen 1997)

In the block universe, every human being is a four-dimensional tube, at the "southern" end of which is his birth and at the "northern" end of which is his death. The thickness of the tube is the three-dimensional extent of the body. This thickness is so small in comparison with the temporal extension that one also speaks of the world-line of the body, thus neglecting the spatial extension. The tube is given as a whole at once. But we experience it as if we were passing through time, moving away from birth and towards death. Hermann Weyl describes it like this:

> *Only from the point of view of the consciousness crawling up in the world lines of the bodies does a section of this world "come to life" and pass it by as a spatial image in temporal transformation.* (Weyl 1948)

This "crawling up of consciousness" is for Weyl something that lies outside the natural scientific description of the world, something that must be added to it.

Does anything change in these ratios when quantum mechanics is added? Maybe (these things are controversial and speculative), but it doesn't look like it's a change for the better. An article on quantum gravity by Hermann Nicolai states:

> The "wave function of the universe" [...] is supposed to contain the complete information about the universe "from beginning to end". A good way to visualise [it] is to think of it as a film reel; "time" and the illusion that "something happens" emerge only when the film is played. (Nicolai 2005)

It is left open who or what plays the film, how and why. Who watches the film? And what is the film reel actually made of? And the projector?

Julian Barbour (2000) takes it even further. He argues that in quantum cosmology causality and temporal sequences disappear completely. There is now only a set of possible moments to which the "wave function of the universe" assigns probabilities, without these moments having any relation to each other. He speaks of an "instantaneous pluralism of moments".

Perhaps one does not even need to know the particular theories and their interpretations to realize that a moving now cannot occur in physics. McTaggart (1908) argues much more generally in his paper *The Unreality of Time*. He distinguishes two time series, which he calls A-series and B-series. In the B-series, there are earlier/later/simultaneous relations between individual moments or events, i.e., one can say how they relate to each other in time. This is the kind of series that physics can describe to us (we will disregard difficulties arising from relativity in this respect, they are not relevant to the argument). The A-series corresponds to the B-series, except that it additionally contains a distinguished moment called "now", which divides time into two halves: on one side the future, on the other the past. Further, McTaggart argues that we can only really call time time when something changes, that is, when the now moves. But what does this mean? In relation to what is it moving? McTaggart tries a few approaches, all of which he leads to a logical contradiction. Finally, he shows: the answer can only be that the now moves with respect to a second time. A particular moment is right now Now, another moment will be Now later, another has been Now earlier. This new Now, indicating where the original Now is right now, must move again in relation to a new, third time, and so on, ad infinitum. So the assumption of a moving now logically leads to an infinite regress. McTaggart concludes that time is logically inconsistent, so it must be an illusion.

In contrast to this, however, one can also take the following standpoint: The moving now is a fundamental part of our experience; it cannot simply be argued away. McTaggart's argument only means that it cannot be fitted into a description of *the* world based on mathematical structures, where time is represented as a *series of instants.*

Lee Smolin (2015) devotes an entire research program ("Temporal Naturalism") to the task of introducing the moving now into physics after all. He sees the logical difficulties, but thinks he might be able to get around them. In particular, he also sees that such physics must not describe the world exclusively as a mathematical structure. For him, naturalism is a prerequisite, that is, the idea that everything is part of a world that exists independently of us and can be deciphered by natural science. Qualia and the strange passing of time, however, are for him also something directly given, which cannot simply be dismissed with psychological explanations. Therefore, for him, these things must be part of the world in some way. Since, according to Smolin, this cannot be achieved in principle with the currently established theories and views, fundamentally new approaches in physics are required. I find the procedure he describes for his research program highly speculative and not necessarily conclusive, but nevertheless wish him every success.

11.3   Qualia and Physicalism

It is of course denied by adherents of physicalism that our subjective experience (qualia) is outside the purview of the natural sciences: Several different views can be distinguished, but we will limit ourselves here to what I consider to be the two most important: eliminationism and functionalism.

Elimination

Qualia do not exist at all. This is what, for example, Daniel Dennett, one of the most influential philosophers of mind, claims. When we think we feel something, we only think it. In the terminology of Sect. 11.1: access consciousness refers to an assumed phenomenal consciousness in much the same way as it refers to an optical illusion. We think and pronounce that there is a red spot on our "inner screen", but in reality there are only information-processing processes in the brain that can **represent** certain aspects of the external world and certain concepts, but not qualia, only the conceptual idea of qualia.

One can argue against this attitude that qualia are completely obvious, more obvious than anything else, that they can be doubted less than anything else, that they are the only thing that is really given to us for sure at a moment; that this attitude is therefore a self-denying ostrich tactic in front of the problem. Yes, the qualia (the experience) are not to be found in the brain, but they are obviously there, which is the problem. Unfortunately, I can't photograph and share them either, so I can't prove the existence of my qualia, but Dennett could be so kind as to look into his own consciousness.

On the other hand, if Dennett happens to be a philosophical zombie, then yes, he would be right. And how would he know that it's different for me? Even if he's not a zombie, why would he "know" anything at all about his own qualia, in the sense of access consciousness? Couldn't it be that some people can't access their qualia at all, so they're "second-order zombies"(my own term) in a sense? And why am I actually so sure that I can't be wrong about my qualia? So, in the end, is Dennett even right? These considerations show how difficult it is to come to definite conclusions on this subject.

I want to emphasize that the elimination approach seems to me logically consistent. It cannot be refuted in the sense of an argument, because I cannot show anyone my qualia, and I cannot show anyone his own qualia. So anyone who denies them is unassailable. But since I am indeed sure that I cannot be mistaken about my qualia, I exclude elimination for myself personally.

Functionalism

Probably the most common attitude among natural scientists towards subjective experience is the view that it is generated by information processing in the brain.

At least since the "cognitive revolution" in the 1950s (i.e. the beginning of strong activity in the field of cognitive and neuroscience, linguistics, computer science and research on artificial intelligence), it has been clear that the brain functions in some ways similar to a computer. Specifically, the brain processes information in digital form. In this context, nerve cells (neurons) can assume exactly two states: transmitting an electrical impulse ("firing") or not ("not firing"). The effect of firing or not firing on neighbouring neurons depends on the type and thickness of the synapses, the connections between neurons. These connections can change over time, explaining memory and learning processes, at least in principle. The brain processes input from the sensory organs by converting it into digital form and directing the information in the form of electrical impulses through the network of neurons in the brain to specific regions where it is further processed by more digital processes. At a higher level (in terms of reductionism; Chap. 5), this means that we recognize and think. Digital processing is so complex that numerous forms and contexts of the external world, as well as higher concepts, are represented in an informatic way. Output, i.e. muscle movement, is also induced in digital form from the brain, especially our linguistic expression, which is also delivered to the outside world through muscle movement.

In the course of these findings functionalism says: For the mental states not all details of the physical states are relevant, but only the information processing matters, the functionality implemented between input and output. Whether this processing is implemented by carbon-based neurons or by anything else is completely irrelevant. That is, if you decouple sensory organs and muscles from the brain and instead connect them to a computer program that functionally does the same thing, i.e. turns the same input into the same output, as the brain (e.g. braking at a red light), and the information processing in between is done in an equivalent way (in the computer science sense), then this program can be said to have the same consciousness as the replaced brain. This has led to science fiction fantasies that we could achieve "immortality" by having our "mind" mapped onto a computer program that is not subject to physical decay like our body.

Functionalism is probably the most common view among natural scientists today. But if you look closely, it is - in contrast to elimination - inconsistent.

In his "Story of a Brain", Zuboff (1981) tells the story of a man in a technologically advanced age who suffers from a serious disease that corrodes his entire body, except for his nervous system. As a result, he has his brain removed and placed in a nutrient solution where it can be kept alive for any length of time. His friends make sure that the brain is connected to certain devices that stimulate neural activation patterns in the brain, which give the man (whose identity is still in the brain) pleasant experiences (because the experience is caused by the consequences of activation patterns that represent the information-processing processes in the brain).

Then a series of events happens that lead to a further and further dissection and transformation of the brain. At each step it is argued that nothing changes in the experiences, the activation patterns are still the same, only their spatial arrangement has changed, but according to functionalism that doesn't matter. First the brain is split into its two halves, but care is taken that the excitations of the neurons at the interfaces are precisely matched as if the halves were still connected. In like manner the brain is divided further and further, until at last each neuron floats in its own nutrient solution. The project is now overseen by all mankind, each taking care of a nerve cell for which he is responsible.

Following a precise plan, everyone knows when to stimulate each cell, so that the overall pattern is the same as if the cells were still connected to each other in the brain. Occasionally cells die and are replaced by new ones, again this should not matter to the experience.

One day, a participant in the experiment accidentally spills the solution of his nerve cell, which was about to be activated. There is not enough time to replace it, and he asks his neighbor for help. The latter replies:

> *Why don't we take my entire bath over to the old position of yours? Then won't it still be the same experience brought about in five minutes that it would have been with the old neuron if we fire this then, since this one is just like the old one? Surely the bath's identity means nothing. Anyway, then we can bring the bath back here and I can use the neuron for the experience it is scheduled to be used for later on. Wait a minute! [...] So why need we move the bath at all? Leave it here; fire it for yours; and then I'll fire it for mine. Both experiences must still come about. Wait a minute again! Then all we need do is fire this one neuron here in place of all the firings of all neurons just like it! Then there need be only one neuron of each type firing again and again and again to bring about all these experiences! But how would the neurons know even that they were repeating an impulse when they fired again and again? How would they know the relative order of their firings? Then we could have one neuron of each sort firing once and that would provide the physical realization of all patterns of impulses. [...] And couldn't these neurons simply be any of those naturally firing in any head? So what are we all doing here? [...] But if all possible neural experience will be brought about simply in the firing once of one of each type of neuron, how can any experiencer believe that he is connected to anything more than this bare minimum of physical reality through the fact of his having any of his experiences? And so all this talk of heads and neurons in them, which is supposedly based on the true discovery of physical realities, is undermined entirely.* (Zuboff 1981)

In *The Mind's I*, Hofstadter and Dennett (1981) attempt to debunk this argument. They claim that if this were true, one could also say that all possible books are realized by printing each letter exactly once. However, this objection is false, and the reason why it is false is very revealing.

The difference with books is that they are read, and they are read in a certain way. We know the decoding rule, i.e. we know that we have to read each line from left to right and the pages from top to bottom. **The point about information is that it only represents something specific if there is an explicit decoding rule for it**. The JPG file represents a picture only because there is an explicit rule about how to convert the ones and zeros into colored pixels. In a text file, the bit sequence 1,000,001 represents the letter A only because the ASCII code says so. A wild jumble of ones and zeros without such a rule represents nothing at all.

In functionalism, however, the ones and zeros in the brain are expected to have meaning on their own, to decode themselves, so to speak, and to produce an experience in the process. This is nonsense, however, as Zuboff so beautifully demonstrates. The brain has its interfaces, where encoding and decoding takes place, on the outside. Incoming sensory data is converted into nerve impulses, outgoing impulses into muscle movements. Only these interfaces define what the ones and zeros "mean" in the brain. All information processing occurs in relation to these interfaces.

If you decouple the external interfaces from the brain and connect them to a computer program instead, then we can say the program is doing the same thing as the brain if the behavior at the interfaces is the same, and "the same" refers **only** to the relationship to those interfaces. Without them, there is no longer a reference point, one can apply arbitrary transformations without changing the "inner" information content, and this leads to the absurd situation in Zuboff's story. The argument is similar in principle to the one I made against the many-worlds interpretation: If a state no longer refers to anything outside of it, then it can no longer be said to have any inner meaning.

Subjective experience is something other than information; information feels nothing. It is also something different from the timeless mathematical structures to which physics reduces everything. Our experience and especially the passing of time do not fit into the picture that physics can draw of reality. They are outside the fundamental limits of physics.

12

Conclusion

We've come to the end of our philosophical/physical underwater journey. I admit it was a lot of difficult, densely packed material. So let's take some time now to review what we've discussed. Above all, I would also like to re-emphasize and summarize some points that are particularly important to me and bring them into a logical context. To do this, we can separate out three main storylines: (1) physics and reality, (2) richness of physics, and (3) practical limits and crisis of physics.

12.1 Physics and Reality

Our starting point was philosophical wonder and questioning. We want to get to the bottom of ourselves and the world, as far as this is possible for us. In doing so, we first encountered **Wittgenstein,** who asserts in the *Tractatus* that the only thing that can really be said are the statements that natural science produces. For only this can express itself clearly; everything else is linguistic confusion, a fog of words. However, he also immediately admits that natural science cannot really explain anything and also does not touch our actual life problems. But one must remain silent about the rest, because nothing can be said about it. Wittgenstein does not allow any grey area here, either something can be said clearly and objectively, or not at all.

We have met **Jaspers** as a kind of indirect opponent. From his point of view, it is not the task of philosophy to say something unambiguous. On the contrary, philosophizing is always a personal spiritual adventure, always subjectively coloured. It is about a truth from which one lives, not about a truth

© Springer-Verlag GmbH Germany, part of Springer Nature 2022
J.-M. Schwindt, *Universe Without Things*,
https://doi.org/10.1007/978-3-662-65426-2_12

that one can prove. Nevertheless, something can be said here, communicated, thoughts can be built upon one another, they are just not as unambiguous as in the natural sciences, but that does not make them any less significant, on the contrary, they go towards a real ground, a realm that natural science can never reach (on this point both agree).

Furthermore, in the course of the book we have discussed **physicalism,** the nowadays widespread view that natural science does get to the bottom of reality, indeed that everything real can be described and explained by the most fundamental of the natural sciences, physics. Basically, large parts of this book can be understood as an attempt to orient ourselves within this triad of Wittgenstein, Jaspers and physicalism.

To this end, we first looked at how natural science actually works, what its characteristics are, how it arrives at its statements and what kind of statements these are. In particular, we have come across a **reductionist hierarchy**, "higher" disciplines or theories that can in *principle* be reduced to "deeper", i.e. more fundamental disciplines or theories. Reduction works first by allowing things to be broken down into their constituent parts and understanding the behavior of the whole thing from the behavior of its constituent parts and their interactions. At the deeper levels of the hierarchical structure, however, it also happens that things are not further decomposed but replaced by something else entirely, a very different kind of reduction. In that case, the objects of the lower level behave in such a way that they create in us a kind of *illusion of* the objects of the higher level.

The lower levels of the hierarchy are occupied by the theories of physics. Thus, the whole human being – as well as all other things in the world - is reduced by the natural sciences to purely physical facts or, even more specifically, according to the current state of knowledge, to the interactions of elementary particles, which are themselves excitations of quantum fields. This is how physicalism comes about: the clarity and high credibility that natural science achieves because of its specific approach, combined with its reductionist character, suggest that everything *is* ultimately nothing but these interacting excitations of quantum fields (or what a hypothetical even lower level of the hierarchy might make of them).

However, in order to better assess this idea, one must look at physics a bit more closely, for it has some traits that are different from the other natural sciences. In my opinion, no one has examined these traits as deeply and accurately as **Duhem** in his work *The Aim and Structure of Physical Theory.* There it is expressed that physical theories are purely mathematical entities, which as such refer to experimental facts in a certain unambiguous way. However, this relation is anything but trivial and often has to be worked out in a complicated, convoluted work of translation. There is no simple one-to-one relationship between the mathematical elements of the theory and the direct observations made in the context of an experiment. Moreover, all experiments have limited precision, and therefore theories are also always approximations and always tentative. Moreover, Duhem emphasizes that physics must beware of metaphysical interpretations of theories, for no agreement can be reached on such an interpretation. Theories do not relate directly to reality, they only reflect certain aspects of reality, in ways that are impossible for us to fathom; it is more of an intuitive truth that we physicists live from, not one that we can prove. We can only prove that a particular theory efficiently summarizes a particular set of experimental facts, within the bounds of its accuracy. In this sense, a physical theory is also not to be understood as an explanation of anything.

Such a view of physics, which as I said I find very plausible, suggests a neat **separation between (1) reality, (2) the world we experience and imagine, and (3) the mathematical structures of physical theories.** Reality underlies everything, but we cannot say anything unambiguous about it. To be more precise: only negative statements can be made about it, it must be *compatible* with the findings of physics, i.e. it must be of such a nature that it *admits* the laws of nature that have been found, and already with this very many statements about reality that have been made in the past can be ruled out. The world experienced by us is the world of appearances which we perceive sensually and which are the basis of our ideas and our view. In this world we also build physical experiments. Physical theories must relate to this world of appearances, must somehow translate into the sensually perceptible, however complicated the experiments and their convoluted interpretations may be that accomplish this. Without this reference, theories are only about **mathematical structures.** By this I mean mathematical sets on which certain relations (or operations, functions, etc.) are defined, as described in Chap. 3.

The separation mentioned is non-physicalist by its nature: the contents of physical theories are separated from reality. However, it also allows for a certain metaphysical speculation, which then amounts to physicalism after all: If nothing definite can be said about reality, at least it is not impossible that it is ultimately identical with the mathematical structures of physical theories (more precisely, with the mathematical structure described by the most fundamental of all theories, the hypothetical theory of everything), that reality is thus a purely **mathematical reality.** This view is held by Max Tegmark, he also elaborated it in quite some detail. The experienced world together with its experiencers, i.e. us, would then have to "emerge" from this mathematical reality. Such a view is in a certain sense very *Platonic:* Everything material is reduced to timeless "pure ideas" (if we characterize the mathematical structures as such, which, however, seems natural).

Quantum mechanics brings some new aspects to these considerations. First of all, it seems to clearly confirm Duhem's view: The mathematical objects of the theory (wavefunctions or state vectors) relate to the experimental facts in an unexpected, complicated way, namely in the form of probability statements, which are, moreover, of such a form that they expose as false many properties that we took for granted in matter. The world of appearances is thus exposed as deceptive. On the other hand, the mathematical side of the theory also seems to be incomplete: it cannot describe what goes on in a measurement and how it comes to the decision between the different possibilities given with the probability statement. The experienced ("classical") world of experimentation and the mathematical structure of the theory both seem to represent only respective incomplete sub-aspects of a broader reality. Physicalism, however, conjures an ace out of its sleeve: the many-worlds interpretation. With this, a purely mathematical-physical reality seems possible again by explaining the decision for one of the possibilities in a measurement as a subjective illusion whose occurrence is a direct consequence of the Schrödinger equation. From my point of view, however, this interpretation has some fundamental problems and is ultimately inconclusive. Thus, for me, QM remains a confirmation of Duhem's view and a strong indication against physicalism and against a purely mathematical reality.

In addition, I have tried to point out **fundamental limits** of physics. In particular, I have given two examples of aspects of reality that we can *recognize* but about which physics has nothing to say. The first example was the passage of time, which does not and cannot occur in the theories of physics because mathematical structures are *per se* timeless. Physics can only introduce a fourth dimension of space that differs from the others via special metric properties and can thereby reproduce some aspects of time, but not its passing. The second example was the "hard problem of consciousness research", i.e. the fact that we subjectively feel and experience something. Both aspects are problematic in their formulation, can't really be defined or explained well. Some people can't make anything out of them at all. Others seem to understand them immediately, are even *sure* that they cannot occur in physics, simply because they are qualitatively different from the mathematical structures from which the latter is constituted. But this difference cannot be shown logically or objectified in the sense of natural science, it can only be "seen" directly or not. Thus, there is an insurmountable gulf between those who "see" the otherness and those who consider it to be a crank, wishful thinking, misunderstanding and illusion.

If you are willing to rule out physicalism with me (whether you do so because you agree with me on the fundamental limits or for other reasons), the question still remains whether we can say or know anything about what goes beyond physics or whether, as Wittgenstein writes in the *Tractatus,* we must remain silent about it. Wittgenstein, Jaspers, and physicalism all agree that beyond the natural sciences and mathematics, no unambiguous statements can emerge that everyone can understand and find to be true. Saying or knowing in this area is thus unlikely to have the character of well-defined, provable propositions. Various alternatives are conceivable: metaphysical speculation, religious faith, philosophical faith (a term defined by Karl Jaspers; we do not have the space or time here to go into it in more detail), poetry, and mystical insight. All of these things can obviously inspire and deeply affect at least some of us, and all of them can also be the subject of some form of communication; we can at least communicate with some about them. But the question remains, in what sense can these things be *valid*, represent a real *truth?* Perhaps it is this last question, which indeed needs to be covered by a cloak of silence.

12.2 The Richness of Physics

We have made a foray through the various fields of physics, listed their most important theories and introduced their central concepts, sometimes rather in hints and metaphors, which remains the only possibility if one (like me in this book) is "forbidden" to use higher mathematics. Thereby we saw, that none of the theories has the truth for itself alone, but that each has essential and surprising things to contribute to our view of the world. Everywhere, relationships and ideas come to light that a philosopher would never have come up with on his own. Only the arduous, testing path of exchange between concept formation and experiment leads to this diversity of perspectives that physics has to offer.

She has amazing things to say about our universe, which in its expansion has evolved from a very hot and dense phase, the Big Bang, to the sparsely populated, largely cold and empty giant we see ourselves surrounded by. There was talk of particles that are not particles at all; of exploding stars in whose heat the matter of which we are made was baked; of a deterministic clockwork universe into which chance then strayed after all; of the question, still not entirely resolved, of what actually happens in a quantum mechanical measurement, what role we as observers play in it; of the time-reversal invariance of the laws of nature; of entropy, which is the only thing that ensures that we can distinguish between past and future; of a curved space in a curved spacetime; of the importance of perspective, of the frame of reference that plays a decisive role in so many measurements.

Physics is the path to a world view that is bursting with creative ideas, versatile, profound, sophisticated, and on top of that, it tells a story based on facts. These facts, however, are difficult to tease apart in detail, because there is so much context, so many presuppositions, and so many nested terminologies attached to all physical statements that it is very difficult to put a concrete situation into sentences that, taken by themselves, express a "physical truth".

This amount of context attached to all statements and the holistic interconnectedness of theories and experiments make the study of physics a very complex, demanding intellectual tinkering that cannot be cut short, but is highly rewarding and literally opens up worlds.

12.3   Practical Limits and Crisis of Physics

Besides the fundamental limits, physics also has **practical limits**. Some areas of physics are not accessible to us because we cannot carry out the experiments and observations necessary for them. Much of the universe is locked behind a cosmic horizon for all time. The high energy scales that interest us most regarding unified theories and quantum gravity are inaccessible to us because, in all likelihood, we will never be able to generate energies of that magnitude. The Big Bang is also beyond our reach; we cannot see what actually happened then and whether there may have been a before.

Thus, it is probable that some of the open questions that today plague the research institutes of physics will never be answered. There has already been a marked slowdown in fundamental discoveries in recent decades; some physicists have been talking about a crisis in fundamental physics for about 20 years now. Since the LHC found nothing new beyond the Standard Model of particle physics, the sense of perplexity has intensified in some areas. I have tried to make it clear that this "crisis" is not primarily the result of misconduct, a wrong focus in research, or physicists "going berserk", as other books claim, but is simply in the nature of things, in the practical limits we face. It is a consequence of the great successes that physics has celebrated over the last few centuries. These successes have led to the fact that most of the phenomena accessible to us are already "covered" by very precise theories.

In *The End of Science,* **John Horgan** argues that the possible insights of science are finite. This does not necessarily refer to the *applications of* the natural sciences; these can possibly be recombined in any number. But that the knowledge itself, the classification of phenomena, the fathoming of cause-effect relationships and their attribution to natural laws, is finite seems to me extremely plausible. After all, it is precisely the strength of natural science, and of physics in particular, to reduce many things to a few. If this "few" would be infinitely extensive in the end, this would make natural science look pretty bad. So it is entirely logical that research in natural science will eventually come to a slowdown and eventually an end (unlike, perhaps, engineering, which continues to generate new applications for all eternity). With the "crisis" in fundamental physics, we are reminded of this finitude. Of course, this end is not necessarily near yet, maybe tomorrow will surprise us with the publication of a groundbreaking new idea that opens up a whole new field of research, and there are still several areas where further discoveries can be expected. But at some point there will be an end. At some point there will be nothing new left for us to find.

But you never know whether you have really reached the end, whether you haven't overlooked anything, whether the next experiment might not reveal a surprise, whether the next flash of inspiration might not bring the theory of everything to light. So even when the end has been reached, the research institutes will continue to exist for quite a while and will continue to safeguard what they have discovered. But at some point, the public's and governments' acceptance of pouring large amounts of money into elaborate experiments will wane. Maybe the whole thing will become a matter of enthusiasts, encouraged by a few private patrons. But maybe at some point the operation will cease, because it's grueling to invest decades in something that doesn't produce anything new.

The question that concerns me is whether there is then a risk that knowledge will be lost. Of course, the areas that are needed for practical applications are not at risk. But much of the physical knowledge that was built up in the twentieth century has purely philosophical relevance; it is crucial to our view of the world, but cannot be applied in engineering terms. This includes cosmology and large parts of particle physics. Will this knowledge still be passed on when there is no more research? People will certainly try to summarize the essentials from these areas and offer them as electives in university courses. But as I said before, a real, deep understanding can only be achieved through many years of work in these fields, and that is best done in the context of research work and in exchange with other active researchers. So there is a real danger that knowledge will be lost.

One hope I have in this regard is the possibility that the immense *personal* spiritual benefits of sustained pursuit in these fields will be recognized, even without producing new research results. I hope, as a small step in this direction, that I have at least succeeded, despite all the dark and confused connections and thoughts I have put on paper here, in highlighting the beauty and value of my great beloved, physics, so that others may also pay homage to it.

Literature

Barbour, J. (2000). *The end of time*. W&N.

Baricco, A. (2003). *Novecento* (4. Aufl.). Piper.

Block, N. (1995). On a confusion about a function of consciousness. *Behavioral and Brain Sciences, 18*, 227.

Cantor, G. (1895). Beiträge zur Begründung der transfiniten Mengenlehre. *Mathematische Annalen, 46*(4), 481.

Carnap, R. (1993). *Mein Weg in die Philosophie*. Reclam..

Carroll, S. (2010). *From eternity to here*. Dutton.

Chalmers. (1995). Facing up to the problem of consciousness. *Journal of Consciousness Studies, 2*(3), 200.

Duhem, P. (1954). *The aim and structure of physical theory*. Princeton University Press.

Dürr, H. -P. (Hrsg.). (1986). *Physik und Transzendenz*. Scherz.

Eddington, A. (1928). *The nature of the physical world*. Macmillian.

Frenkel, E. (2013). *Love and Math*. Basic Books.

Frisch, M. (1964). *Mein Name sei Gantenbein*. Suhrkamp..

Hawking, S. (1988). *A brief history of time*. Bantam Books.

Helling, R. (2011). *How I learned to stop worrying and love QFT*. https://arxiv.org/pdf/1201.2714.pdf

Hirschberger, J. (1980). *Geschichte der Philosophie* (12. Aufl.). Herder.

Hofstadter, D. R., & Dennett, D. C. (1981). *The Mind's I*. Basic Books.

Horgan, J. (1996). *The end of science*. Addison Wesley.

Hossenfelder, S. (2018). *Lost in math*. Basic Books.

Hut, P., & van Fraassen, B. (1997). Elements of reality. *Journal of Consciousness Studies, 4*(2), 167.

Jackson, F. (1986). What Mary didn't know. *Journal of Philosophy, 83*, 291.

Jaspers, K. (1960). *Vernunft und Freiheit*. Europ. Buchklub.

Kant, I. (1998). *Critique of pure reason*. Cambridge University Press.

© Springer-Verlag GmbH Germany, part of Springer Nature 2022
J.-M. Schwindt, *Universe Without Things*,
https://doi.org/10.1007/978-3-662-65426-2

Mach, E. (1882). *Die ökonomische Natur der physikalischen Forschung.* Akademie der Wissenschaften. (Vortrag an der kaiserl. Akademie der Wissenschaften, Wien).

Marias, J. (1995). *A heart so white.* Harvill Press.

McTaggart. (1908). The unreality of time. *Mind: A Quarterly Review of Psychology and Philosophy, 17,* 456.

Musil, R. (1952). *Der Mann ohne Eigenschaften.* Rowohlt.

Nagel, T. (1974). What is it like to be a bat? *Philosophy Reviews, 83*(4), 435.

Nicolai, H. (2005). Loop quantum gravity: An outside view. *Classical and Quantum Gravity, 22,* R193.

Schlosshauer, M. (Hrsg.). (2011). *Elegance and Enigma – The quantum interviews.* Springer.

Schrödinger, E. (1958). *Geist und Materie.* Paul Zsolnay.

Schwarzschild K. (1916). *Über das Gravitationsfeld eines Massenpunktes nach der Einsteinschen Theorie,* Sitzungsberichte d. Königlich Preußischen Akademie der Wissenschaften (Berlin) vom 3.2.1916, S. 189.

Schwindt, J. -M. (2012). *Nothing happens in the Universe of the Everett Interpretation.* https://arxiv.org/pdf/1210.8447.pdf

Smolin, L. (2006). *The trouble with physics.* Houghton Mifflin Harcourt.

Smolin, L. (2015). Temporal naturalism. *Studies in History and Philosophy of Modern Physics, 52A,* 86.

Tegmark, M. (2014). *Our mathematical universe.* Vintage Books.

Unzicker, A. (2010). *Vom Urknall zum Durchknall.* Springer.

Villani, C. (2015). *Birth of a theorem.* Farrar, Straus and Giroux.

Weinberg, S. (1977). *The first three minutes.* Basic Books.

Wetterich, C. (2013). A universe without expansion. *Physics of the Dark Universe, 2*(4), 184. https://arxiv.org/pdf/1303.6878.pdf

Weyl, H. (1948). *Philosophie der Mathematik und Naturwissenschaft.* Oldenbourg.

Wittgenstein, L. (1963). *Tractatus logico-philosophicus.* Suhrkamp.

Woit, P. (2006). *Not even wrong.* Basic Books.

Zeh, H. (1989). *The physical basis of the direction of time.* Springer.

Zuboff, A. (1981). *Story of a Brain.* In: Hofstadter & Dennett (1981).

Index

Printed in the United States
by Baker & Taylor Publisher Services